ELEMENTS IN SPECTROSCOPY

ELEMENTS IN SPECTROSCOPY

ELEMENTS IN SPECTROSCOPY

Owen Parkers

Oxford Book Company

Jaipur ▪ New Delhi

Published By

Oxford Book Company

Regd. Off. "Bony Residency", Gate No. 2
Opp. Tilak Public School, Vishweswariya Nagar,
Gopalpura Road, Jaipur - 302018, India
Ph: 0141 - 2761280, Telefax: 2761381
email - oxfordbookjpr@gmail.com
website - oxfordbookcompany.com

Delhi Office: 102, 1st Floor, Satyam House,
4327/3, Ansari Road, Darya Ganj,
New Delhi - 110002, India
Ph: 011-45652440

Distributed By
KEYA INTERNATIONAL PRESS LLP
E-mail: keyainternationalpressllp@gmail.com

Published in India, 2025

ISBN : 9789391096984

Typeset by Shivangi Computers, Jaipur

Printed and bound in India by Thomson Press Ltd.,
New Delhi

Contents

Preface

Spectroscopy pertains to the dispersion of an object's light into its component colors (i.e. energies). By performing this dissection and analysis of an object's light, astronomers can infer the physical properties of that object (such as temperature, mass, luminosity and composition). But before we hurtle headlong into the wild and woolly field of spectroscopy, we need to try to answer some seemingly simple questions, such as *what is light*? And *how does it behave*? These questions may seem simple to you, but they have presented some of the most difficult conceptual challenges in the long history of physics. It has only been in this century, with the creation of quantum mechanics that we have gained a quantitative understanding of how light and atoms work. You see, the questions we pose are not always easy, but to understand and solve them will unlock a new way of looking at our Universe.

Typically one can observe two distinctive classes of spectra: *continous* and *discrete*. For a continuous spectrum, the light is composed of a wide, continuous range of colors (energies). With discrete spectra, one sees only bright or dark lines at very distinct and sharply-defined colors (energies). As we'll discover shortly, discrete spectra with bright lines are called *emission* spectra, those with dark lines are termed *absorption* spectra. Continuous spectra arise from dense gases or solid objects which radiate their heat away through the production of light. Such objects emit light over a broad range of wavelengths, thus the apparent spectrum seems smooth and continuous. Stars emit light in a *predominantly* (but not completely!) continuous spectrum. Other examples of such objects are incandescent light bulbs, electric cooking stove burners, flames, cooling fire embers and...you. Yes, you, right this minute, are emitting a continuous spectrum — but the light waves you're emitting are not visible — they lie at infrared wavelengths (i.e. lower energies, and longer wavelengths than even red light). Unlike a continuous spectrum source, which can have *any* energy it wants (all you have to do is change the temperature), the electron clouds surrounding the nuclei of atoms can have only very specific energies dictated by quantum mechanics. Each element on the periodic table has its own set of possible energy levels, and with few exceptions the levels are distinct and identifiable.

Atoms will also tend to settle to the lowest energy level (in spectroscopist's lingo, this is called the *ground state*). This means that an excited atom in a higher energy level must 'dump' some energy. The way an atom 'dumps' that energy is by emitting a wave of light with that *exact* energy. Tiny changes of energy in an atom generate photons with small energies and long wavelengths, such as radio waves! Similarly, large changes of energy in an atom will mean that high-energy, short-wavelength photons (UV, x-ray, gamma-rays) are emitted. On the other hand, what would happen if we tried to reverse this process? That is, what would happen if we fired this special photon back into a ground state atom? That's right, the atom could absorb that 'specially-energetic' photon and would become excited, jumping from the ground state to a higher energy level. If a star with a 'continuous' spectrum is shining upon an atom, the wavelengths corresponding to possible energy transitions within that atom will be absorbed and therefore an observer will not see them.

These book are devised for such courses. Separate chapters are devoted for the various analytical techniques.

—Owen Parkers

Preface

Spectroscopy pertains to the dispersion of an object's light into its component colors (i.e. energies). By performing this dissection and analysis of an object's light, astronomers can infer the physical properties of that object (such as temperature, mass, luminosity and composition). But before we hurtle headlong into the wild and woolly field of spectroscopy, we need to try to answer some seemingly simple questions, such as *what is light*? And *how does it behave*? These questions may seem simple to you, but they have presented some of the most difficult conceptual challenges in the long history of physics. It has only been in this century, with the creation of quantum mechanics that we have gained a quantitative understanding of how light and atoms work. You see, the questions we pose are not always easy, but to understand and solve them will unlock a new way of looking at our Universe.

Typically one can observe two distinctive classes of spectra: *continous* and *discrete*. For a continuous spectrum, the light is composed of a wide, continuous range of colors (energies). With discrete spectra, one sees only bright or dark lines at very distinct and sharply-defined colors (energies). As we'll discover shortly, discrete spectra with bright lines are called *emission* spectra, those with dark lines are termed *absorption* spectra. Continuous spectra arise from dense gases or solid objects which radiate their heat away through the production of light. Such objects emit light over a broad range of wavelengths, thus the apparent spectrum seems smooth and continuous. Stars emit light in a *predominantly* (but not completely!) continuous spectrum. Other examples of such objects are incandescent light bulbs, electric cooking stove burners, flames, cooling fire embers and...you. Yes, you, right this minute, are emitting a continuous spectrum — but the light waves you're emitting are not visible — they lie at infrared wavelengths (i.e. lower energies, and longer wavelengths than even red light). Unlike a continuous spectrum source, which can have *any* energy it wants (all you have to do is change the temperature), the electron clouds surrounding the nuclei of atoms can have only very specific energies dictated by quantum mechanics. Each element on the periodic table has its own set of possible energy levels, and with few exceptions the levels are distinct and identifiable.

Atoms will also tend to settle to the lowest energy level (in spectroscopist's lingo, this is called the *ground state*). This means that an excited atom in a higher energy level must 'dump' some energy. The way an atom 'dumps' that energy is by emitting a wave of light with that *exact* energy. Tiny changes of energy in an atom generate photons with small energies and long wavelengths, such as radio waves! Similarly, large changes of energy in an atom will mean that high-energy, short-wavelength photons (UV, x-ray, gamma-rays) are emitted. On the other hand, what would happen if we tried to reverse this process? That is, what would happen if we fired this special photon back into a ground state atom? That's right, the atom could absorb that 'specially-energetic' photon and would become excited, jumping from the ground state to a higher energy level. If a star with a 'continuous' spectrum is shining upon an atom, the wavelengths corresponding to possible energy transitions within that atom will be absorbed and therefore an observer will not see them.

These book are devised for such courses. Separate chapters are devoted for the various analytical techniques.

—Owen Parkers

Chapter 1

Spectroscopy

Spectroscopy is the study of the interaction between matter and radiated energy. Historically, spectroscopy originated through the study of visible light dispersed according to its wavelength, e.g., by a prism. Later the concept was expanded greatly to comprise any interaction with radiative energy as a function of its wavelength or frequency. Spectroscopic data is often represented by a spectrum, a plot of the response of interest as a function of wavelength or frequency. Spectrometry and spectrography are terms used to refer to the measurement of radiation intensity as a function of wavelength and are often used to describe experimental spectroscopic methods. Spectral measurement devices are referred to as spectrometres, spectrophotometres, spectrographs or spectral analysers.

Daily observations of colour can be related to spectroscopy. Neon lighting is a direct application of atomic spectroscopy. Neon and other noble gases have characteristic emission colours, and neon lamps use electricity to excite these emissions. Inks, dyes and paints include chemical compounds selected for their spectral characteristics in order to generate specific colours and hues. A commonly encountered molecular spectrum is that of nitrogen dioxide. Gaseous nitrogen dioxide has a characteristic red absorption feature, and this gives air polluted with nitrogen dioxide a reddish brown colour. Rayleigh scattering is a spectroscopic scattering phenomenon that accounts for the colour of the sky.

Spectroscopic studies were central to the development of quantum mechanics and included Max Planck's explanation of blackbody radiation, Albert Einstein's explanation of the photoelectric effect and Niels Bohr's explanation of atomic structure and spectra. Spectroscopy is used in physical and analytical chemistry because atoms and molecules have unique spectra. These spectra can be interpreted to derive information about the atoms and molecules, and they can also be used to detect, identify and quantify chemicals. Spectroscopy is also used in astronomy and remote sensing. Most research telescopes have spectrographs. The measured spectra are used to determine the chemical composition and physical properties of astronomical objects (such as their temperature and velocity).

Spectroscopy is a very important tool in astronomy. It is detailed study of the light from an object. Light is energy that moves through space and can be thought of as either waves or particles. The distances between the peaks of the waves of light are called the light's wavelength. Light is made up of many different wavelengths. For example, visible light has wavelengths of about 1/10th of a micrometre - ten thousand wavelengths would be the width of a dime.

Spectrometres are instruments which spread light out into its wavelengths creating a spectra. Within this spectra, astronomers can study emission and absorption lines which are the fingerprints of atoms and molecules.

An emission line occurs when an electron drops down to a lower orbit around the nucleus of an atom and looses energy. An absorption line occurs when electrons move to a higher orbit by absorbing energy. Each atom has a unique spacing of orbits and can emit or absorb only certain energies or wavelengths.

This is why the location and spacing of spectral lines is unique for each atom. Astronomers can learn a great deal about an object in space by studying its spectrum, such as it's composition (what its made of), temperature, density, and it's motion (both it's rotation as well as how fast it is moving towards or away from us). There are three types of spectra which an object can emit: continuous, emission and absorption spectra. The examples of these types of spectra shown below are for visible light as it is spread out from purple to red, but the concept is the same for any region of the electromagnetic spectrum.

Continuous Spectra

Continuous spectra (also called a thermal or blackbody spectra) are emitted by any object that radiates heat (has a temperature). The light is spread out into a continuous band with every wavelength having some amount of radiation. For example, when sunlight is passed through a prism, it's light is spread out into it's colours.

Absorption Spectraok

The absorption spectra of hydrogen - can you see this pattern in the solar spectrum above this image? Hint: hydrogen is the most abundant element in the sun - look at the darkest lines.

Emission Spectra

An emission spectra occurs when the atoms and molecules in a hot gas emit extra light at certain wavelengths, causing bright lines to appear in a spectra. As with absorption spectra, the pattern of these lines are unique for each element. We can see emission spectra from comets, nebula and certain types of stars.

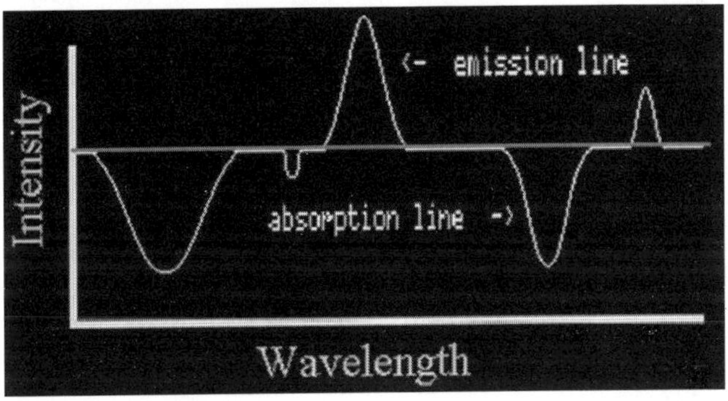

In practice, astronomers rarely look at spectra the way they are displayed in the above images. Instead they study plots of intensity, signal or flux versus wavelength. These plots show how much light is present or absent at each wavelength. A peak in the plot shows the position of an emission line and dip shows where an absorption line is. The spacing and location of these lines are unique to each atom and molecule.

The shape of the continuous spectra (often referred to as the continuum) on a plot is dependent on temperature and motion of the emitting gas. In this simple plot it is shown as a flat line - in reality it is usually a curved line. Also, many of the real data plots you will see have the wavelength or frequency on a logarithmic scale.

Theory

One of the central concepts in spectroscopy is a resonance and its corresponding resonant frequency. Resonances were first characterised in mechanical systems such as pendulums. Mechanical systems that vibrate or oscillate will experience large amplitude oscillations when they are driven at their resonant frequency. A plot of amplitude vs. excitation frequency will have a peak centred at the resonance frequency. This plot is one type of spectrum, with the peak often referred to as a spectral line, and most spectral lines have a similar appearance.

In quantum mechanical systems, the analogous resonance is a coupling of two quantum mechanical stationary states of one system, such as an atom, via an oscillatory source of energy such as a photon. The coupling of the two states is strongest when the energy of the source matches the energy difference between the two states. The energy (E) of a photon is related to its frequency (v) by $E=hv$ where h is Planck's constant, and so a spectrum of the system response vs. photon frequency will peak at the resonant frequency or energy. Particles such as electrons and neutrons have a comparable relationship, the de Broglie relations, between their kinetic energy and their wavelength and frequency and therefore can also excite resonant interactions.

Spectra of atoms and molecules often consist of a series of spectral lines, each one representing a resonance between two different quantum states.

The explanation of these series, and the spectral patterns associated with them, were one of the experimental enigmas that drove the development and acceptance of quantum mechanics. The hydrogen spectral series in particular was first successfully explained by the Rutherford-Bohr quantum model of the hydrogen atom. In some cases spectral lines are well separated and distinguishable, but spectral lines can also overlap and appear to be a single transition if the density of energy states is high enough.

Classification of Methods

Spectroscopy is a sufficiently broad field that many sub-disciplines exist, each with numerous implementations of specific spectroscopic techniques. The various implementations and techniques can be classified in several ways.

Type of Radiative Energy

Types of spectroscopy are distinguished by the type of radiative energy involved in the interaction. In many applications, the spectrum is determined by measuring changes in the intensity or frequency of this energy. The types of radiative energy studied include:

- Electromagnetic radiation was the first source of energy used for spectroscopic studies. Techniques that employ electromagnetic radiation are typically classified by the wavelength region of the spectrum and include microwave, terahertz, infrared, near infrared, visible and ultraviolet, x-ray and gamma spectroscopy.
- Particles, due to their de Broglie wavelength, can also be a source of radiative energy and both electrons and neutrons are commonly used. For a particle, its kinetic energy determines its wavelength.
- Acoustic spectroscopy involves radiated pressure waves.
- Mechanical methods can be employed to impart radiating energy, similar to acoustic waves, to solid materials.

Nature of the Interaction

Types of spectroscopy can also be distinguished by the nature of the interaction between the energy and the material. These interactions include:

- Absorption occurs when energy from the radiative source is absorbed by the material. Absorption is often determined by measuring the fraction of energy transmitted through the material; absorption will decrease the transmitted portion.
- Emission indicates that radiative energy is released by the material. A material's blackbody spectrum is a spontaneous emission spectrum determined by its temperature. Emission can also be induced by other sources of energy such as a flames or sparks or electromagnetic radiation in the case of fluorescence.
- Elastic scattering and reflection spectroscopy determine how incident radiation is reflected or scattered by a material. Crystallography employs the scattering of high energy radiation, such as x-rays and electrons, to examine the arrangement of atoms in proteins and solid crystals.
- Impedance spectroscopy studies the ability of a medium to impede or slow the transmittance of energy. For optical applications, this is characterised by the index of refraction.
- Inelastic scattering phenomena involve an exchange of energy between the radiation and the matter that shifts the wavelength

of the scattered radiation. These include Raman and Compton scattering.

- Coherent or resonance spectroscopy are techniques where the radiative energy couples two quantum states of the material in a coherent interaction that is sustained by the radiating field. The coherence can be disrupted by other interactions, such as particle collisions and energy transfer, and so often require high intensity radiation to be sustained. Nuclear magnetic resonance (NMR) spectroscopy is a widely used resonance method and ultrafast laser methods are also now possible in the infrared and visible spectral regions.

Type of Material

Spectroscopic studies are designed so that the radiant energy interacts with specific types of matter.

Atoms

Atomic spectroscopy was the first application of spectroscopy developed. Atomic absorption spectroscopy (AAS) and atomic emission spectroscopy (AES) involve visible and ultraviolet light. These absorptions and emissions, often referred to as atomic spectral lines, are due to electronic transitions of an outer shell electron to an excited state. Atoms also have distinct x-ray spectra that are attributable to the excitation of inner shell electrons to excited states.

Atoms of different elements have distinct spectra and therefore atomic spectroscopy allows for the identification and quantitation of a sample's elemental composition. Robert Bunsen, developer of the Bunsen burner, and Gustav Kirchhoff discovered new elements by observing their emission spectra. Atomic absorption lines are observed in the solar spectrum and referred to as Fraunhofer lines after their discoverer. A comprehensive explanation of the hydrogen spectrum was an early success of quantum mechanics and explaining the Lamb shift observed in the hydrogen spectrum led to the development of quantum electrodynamics.

Modern implementations of atomic spectroscopy for studying visible and ultraviolet transitions include flame emission spectroscopy, inductively coupled plasma atomic emission spectroscopy, glow discharge spectroscopy, microwave induced plasma spectroscopy, and spark or arc emission spectroscopy. Techniques for studying x-ray spectra include X-ray spectroscopy and X-ray fluorescence (XRF).

Molecules

The combination of atoms into molecules leads to the creation of unique types of energetic states and therefore unique spectra of the transitions between these states. Molecular spectra can be obtained due to electron spin states (electron paramagnetic resonance), molecular rotations, molecular vibration and electronic states. Rotations are collective motions of the atomic nuclei and typically lead to spectra in the microwave and millimetre-wave spectral regions; rotational spectroscopy and microwave spectroscopy are synonymous. Vibrations are relative motions of the atomic nuclei and are studied by both infrared and Raman spectroscopy. Electronic excitations are studied using visible and ultraviolet spectroscopy as well as fluorescence spectroscopy. Studies in molecular spectroscopy led to the development of the first maser and contributed to the subsequent development of the laser.

Crystals and Extended Materials

The combination of atoms or molecules into crystals or other extended forms leads to the creation of additional energetic states. These states are numerous and therefore have a high density of states. This high density often makes the spectra weaker and less distinct, i.e., broader. For instance, blackbody radiation is due to the thermal motions of atoms and molecules within a material. Acoustic and mechanical responses are due to collective motions as well.

Pure crystals, though, can have distinct spectral transitions and the crystal arrangement also has an effect on the observed molecular spectra. The regular lattice structure of crystals also scatters x-rays, electrons or neutrons allowing for crystallographic studies.

Nuclei

Nuclei also have distinct energy states that are widely separated and lead to gamma ray spectra. Distinct nuclear spin states can have their energy separated by a magnetic field, and this allows for NMR spectroscopy.

Other Types

Other types of spectroscopy are distinguished by specific applications or implementations:

- Auger spectroscopy is a method used to study surfaces of materials on a micro-scale. It is often used in connection with electron microscopy.

- Cavity ring down spectroscopy
- Circular Dichroism spectroscopy
- Coherent anti-Stokes Raman spectroscopy (CARS) is a recent technique that has high sensitivity and powerful applications for *in vivo* spectroscopy and imaging.
- Cold vapour atomic fluorescence spectroscopy
- Correlation spectroscopy encompasses several types of two-dimensional NMR spectroscopy.
- Deep-level transient spectroscopy measures concentration and analyses parameters of electrically active defects in semiconducting materials
- Dual polarisation interferometry measures the real and imaginary components of the complex refractive index
- EPR spectroscopy
- Force spectroscopy
- Fourier transform spectroscopy is an efficient method for processing spectra data obtained using interferometres. Fourier transform infrared spectroscopy (FTIR) is a common implementation of infrared spectroscopy. NMR also employs Fourier transforms.
- Hadron spectroscopy studies the energy/mass spectrum of hadrons according to spin, parity, and other particle properties. Baryon spectroscopy and meson spectroscopy are both types of hadron spectroscopy.
- Hyperspectral imaging is a method to create a complete picture of the environment or various objects, each pixel containing a full visible, VNIR, NIR, or infrared spectrum.
- Inelastic electron tunnelling spectroscopy (IETS) uses the changes in current due to inelastic electron-vibration interaction at specific energies that can also measure optically forbidden transitions.
- Inelastic neutron scattering is similar to Raman spectroscopy, but uses neutrons instead of photons.
- Laser-Induced Breakdown Spectroscopy (LIBS), also called Laser-induced plasma spectrometry (LIPS)
- Laser spectroscopy uses tunable lasers and other types of

coherent emission sources, such as optical parametric oscillators, for selective excitation of atomic or molecular species.

- Mass spectroscopy is an historical term used to refer to mass spectrometry. Current recommendations are to use the latter term. Use of the term mass spectroscopy originated in the use of phosphor screens to detect ions.
- Mössbauer spectroscopy probes the properties of specific isotopic nuclei in different atomic environments by analysing the resonant absorption of gamma-rays.
- Neutron spin echo spectroscopy measures internal dynamics in proteins and other soft matter systems
- Photoacoustic spectroscopy measures the sound waves produced upon the absorption of radiation.
- Photoemission spectroscopy
- Photothermal spectroscopy measures heat evolved upon absorption of radiation.
- Raman optical activity spectroscopy exploits Raman scattering and optical activity effects to reveal detailed information on chiral centres in molecules.
- Raman spectroscopy
- Saturated spectroscopy
- Scanning tunneling spectroscopy
- Spectrophotometry
- Time-resolved spectroscopy measures the fluorescence/ phosphorescence lifetime decay rate(s) with respect to emission wavelength.
- Time-Stretch Spectroscopy
- Thermal infrared spectroscopy measures thermal radiation emitted from materials and surfaces and is used to determine the type of bonds present in a sample as well as their lattice environment. The techniques are widely used by organic chemists, mineralogists, and planetary scientists.
- Ultraviolet photoelectron spectroscopy (UPS)
- Vibrational circular dichroism spectroscopy
- X-ray photoelectron spectroscopy (XPS)

Applications:

- Estimate weathered wood exposure times using near infrared spectroscopy.
- Cure monitoring of composites using Optical fibres

Astronomical Spectroscopy

Astronomical spectroscopy is the technique of spectroscopy used in astronomy. The object of study is the spectrum of electromagnetic radiation, including visible light, which radiates from stars and other celestial objects. Spectroscopy can be used to derive many properties of distant stars and galaxies, such as their chemical composition, but also their motion by Doppler shift measurements.

Stars

Astronomical spectroscopy began with Isaac Newton's initial observations of the light of the Sun, dispersed by a prism. He saw a rainbow of colour, and may even have seen absorption lines. These dark bands which appear throughout the solar spectrum were first described in detail by Joseph von Fraunhofer. Most stellar spectra share these two dominant features of the Sun's spectrum: emission at all wavelengths across the optical spectrum (the *continuum*) with many discrete absorption lines, resulting from gaps of radiation.

Fraunhofer's original (1817) designations of absorption lines in the solar spectrum

Letter	*Wavelength (nm)*	*Chemical origin*	*Colour range*
A	759.37	atmospheric O_2	dark red
B	686.72	atmospheric O_2	red
C	656.28	hydrogen alpha	red
D1	589.59	neutral sodium	red orange
D2	589.00	neutral sodium	yellow
E	526.96	neutral iron	green
F	486.13	hydrogen beta	cyan
G	431.42	CH molecule	blue
H	396.85	ionised calcium	dark violet
K	393.37	ionised calcium	dark violet

Fraunhofer and Angelo Secchi were among the pioneers of spectroscopy of the Sun and other stars. Secchi is particularly noted for classifying stars into spectral types, based on the number and strength of the absorption lines in their spectra. Later the origin of the spectral types was found to be related to the temperature of the surface of the star: particular absorption lines can be observed only for a certain range of temperatures; because only in that range are the involved atomic energy levels populated.

The absorption lines in stellar spectra can be used to determine the chemical composition of the star. Each element is responsible for a different set of absorption lines in the spectrum, at wavelengths which can be measured extremely accurately by laboratory experiments. Then, an absorption line at the given wavelength in a stellar spectrum shows that the element must be present. Of particular importance are the absorption lines of hydrogen (which is found in the atmosphere of nearly every star); the hydrogen lines within the visual spectrum are known as Balmer lines.

In 1868, Sir Norman Lockyer observed strong yellow lines in the solar spectrum which had never been seen in laboratory experiments. He deduced that they must be due to an unknown element, which he called helium, from the Greek helios (sun). Helium wasn't conclusively detected on earth until 25 years later.

Also in the 1860s, emission lines (particularly a green line) were observed in the coronal spectrum during solar eclipses that did not correspond to any known spectral lines. Again it was proposed that these were due to an unknown element, provisionally named coronium. It was not until the 1930s that it was discovered that these lines were due to highly ionised iron and nickel, the high ionisation being due to the extreme temperature of the solar corona.

In conjunction with atomic physics and models of stellar evolution, stellar spectroscopy is today used to determine a multitude of properties of stars: their distance, age, luminosity and rate of mass loss can all be estimated from spectral studies, and Doppler shift studies can uncover the presence of hidden companions such as black holes and exoplanets.

Nebulae

In the early days of telescopic astronomy, the word nebula was used to describe any fuzzy patch of light that didn't look like a star. Many of these, such as the Andromeda Nebula, had spectra that looked in many ways a lot like stellar spectra, and these turned out to be galaxies. Others, such as the Cat's Eye Nebula, had very different spectra. When William Huggins looked at the Cat's Eye, he found no continuous spectrum like that seen in the Sun, but just a few strong emission lines. These lines did not correspond to any known elements on earth, and so just as helium had been identified in the Sun, astronomers suggested that the lines were due to a new element, nebulium (occasionally *nebulum* or *nephelium*).

The hypothetical nebulium that was invoked to account for certain bright lines in gaseous nebulae were shown by Ira Sprague Bowen in 1927 as due to doubly ionized oxygen at extremely low density. As Henry Norris Russell put it, "Nebulium has vanished into thin air." But nebulae are typically extremely rarefied, much less dense than the hardest vacuum ever produced on earth. In these conditions, atoms behave quite differently and lines can form which are suppressed at normal densities. These lines are known as forbidden lines, and are the strongest lines in most nebular spectra.

Galaxies

The spectra of galaxies look somewhat similar to stellar spectra, as they consist of the light from millions of stars combined. Galactic spectroscopy has led to many fundamental discoveries. Edwin Hubble discovered in the 1920s that, apart from the nearest ones (those in what is known as the Local Group), all galaxies are receding from the Earth. The further away a galaxy, the faster it is receding. This was the first indication that the universe originated from a single point, in a Big Bang. Doppler shift studies of clusters of galaxies by Fritz Zwicky found that most galaxies were moving much faster than seemed to be possible, from what was known about the mass of the cluster. Zwicky hypothesised that there must be a great deal of non-luminous matter in the galaxy clusters, which became known as dark matter.

Quasars

In the 1950s, some strong radio sources were found to be associated with very dim objects that seemed to be very blue. These were named *Quasi-stellar radio sources*, or quasars. When the first spectrum of one of these objects was taken, it was something of a mystery, with absorption lines at wavelengths where none were expected. It was soon realised that what was being seen was a normal galactic spectrum, but highly redshifted. According to Hubble's Law, this implied that the quasar must be extremely distant, and therefore highly luminous. Quasars are now thought to be galaxies forming, with their extreme energy output being powered by super-massive black holes.

Planets and Asteroids

Planets and asteroids shine only by reflecting the light of their parent star. The reflected light contains absorption bands due to minerals in the rocks present for rocky bodies, or due to the elements and molecules present in the atmospheres of the Gas giants. Asteroids can be classified into three main types, according to their spectra: the C-types are made of carbonaceous material, S-types consist mainly of silicates, and M-types are 'metallic'. C- and S-type asteroids are the most common.

Comets

The spectra of comets consist of a reflected solar spectrum from the dusty clouds surrounding the comet, as well as emission lines from gaseous atoms and molecules excited by sunlight fluorescence and/or chemical reactions.

Nearby comets can even be seen in X-ray as solar wind ions flying to the coma are neutralized, and cometary X-ray spectra therefore reflect the state of the solar wind rather than that of the comet. Many organic chemicals are known to exist in comets, and it has been suggested that cometary impacts provided the Earth with much of the water for its oceans and the chemicals necessary for the formation of life. It has even been suggested that life may have been brought to earth from interstellar space by comets (the Panspermia theory).

Mechanisms

Absorption spectra of nebulae and planetary atmospheres (and gases in terrestrial experiments) arise simply when molecules absorb light of certain frequencies and re-emit it, if at all, in directions away from the observer.

Absorption spectra of stars arise from a different mechanism. Starlight at most frequencies is emitted from close enough to the surface that it can escape; this outer layer is called the photosphere. At the frequencies that molecules, atoms, or ions emit and absorb strongly, most light from deeper parts of the photosphere is absorbed before escaping. The light we see at those absorbed frequencies comes from still shallower depths. The shallow regions are cooler, so the light coming from them is weaker. Thus the spectrum contains relatively dark bands at the characteristic frequencies of constituents of the outer layers.

Amateur Spectroscopy

It is not hard to view spectra of astronomical objects. A simple, homemade DVD spectrograph can be used to view solar spectra. Note that looking at the Sun directly or by reflection can result in temporary or permanent loss of vision. Proper equipment must be used, and children should be supervised. Recently, there has been a resurgence in amateur astronomical spectroscopy. It is possible to use a digital camera or telescope to easily view spectra of astronomical objects. Inexpensive diffraction gratings like the Paton Hawksley Star Analyser or the Rainbow Optics Star Spectroscope can be used to split the starlight. Software like RSpec can be used to plot the curves. An excellent introduction to amateur spectroscopy is *Astronomical Spectroscopy for Amateurs*, published in 2011. Also, *Sky & Telescope Magazine* has produced an excellent video interview that explains how to get started.

Applied Spectroscopy

Applied spectroscopy is the application of various spectroscopic methods for detection and identification of different elements/ compounds in solving problems in the fields of forensics, medicine, oil industry, atmospheric chemistry, pharmacology, etc.

Spectroscopic Methods

A common spectroscopic method for analysis is FTIR spectroscopy, where chemical bonds can be detected through their characteristic infra-red absorption frequencies or wavelengths. These absorption characteristics make infrared analysers an invaluable tool in geoscience, environmental science and atmospheric science. For instance, atmospheric gas monitoring has been facilitated by the development of commercially available gas analysers which can distinguish between carbon dioxide, methane, carbon monoxide, oxygen and nitric oxide. UV spectroscopy is used where strong absorption of ultra-violet radiation occurs in a substance. Such groups are known as chromophores and include aromatic groups, conjugated system of bonds, carbonyl groups and so on. NMR spectroscopy detects hydrogen atoms in specific environments, and complements both IR and UV spectroscopy. The use of Raman spectroscopy is growing for more specialist applications. There are also derivative methods such as infrared microscopy, which allows very small areas to be analysed in an optical microscope. One method of elemental analysis that is important in forensic analysis is EDX performed in the environmental scanning electron microscope, or ESEM. The method involves analysis of back-scattered X-rays from the sample as a result of interaction with the electron beam. Automated SEM-EDS analysis is further used in a range of Automated Mineralogy quantitative mineral, identification and textural mapping, such as the QEMSCAN and MLA solutions.

Sample Preparation

In all three spectroscopic methods, the sample usually needs to be present in solution, which may present problems during forensic examination because it necessarily involves sampling solid from the object to be examined.

FTIR: Three types of samples can be analysed, a solution (KBr), a powder, or a film. A solid film is the easiest and most straight forward sample type to test.

Analysis of Polymers

Many polymer degradation mechanisms can be followed using infra-red spectroscopy, such as UV degradation and oxidation, among many other failure modes.

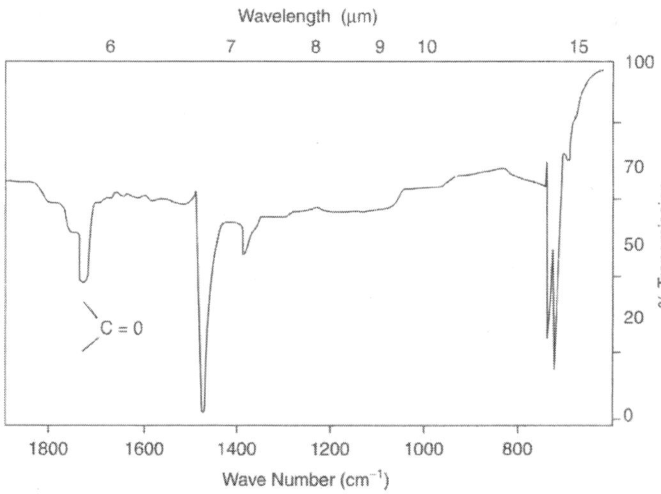

Figure: IR spectrum showing carbonyl absorption due to UV degradation of polyethylene

UV Degradation

Many polymers are attacked by UV radiation at vulnerable points in their chain structures. Thus, polypropylene suffers severe cracking in sunlight unless anti-oxidants are added. The point of attack occurs at the tertiary carbon atom present in every repeat unit, causing oxidation and finally chain breakage. Polyethylene is also susceptible to UV degradation, especially those variants that are branched polymers such as LDPE. The branch points are tertiary carbon atoms, so polymer degradation starts there and results in chain cleavage, and embrittlement. In the example shown at left, carbonyl groups were readily detected by IR spectroscopy from a cast thin film. The product was a road cone that had cracked in service, and many similar cones also failed because an anti-UV additive had not been used.

Oxidation

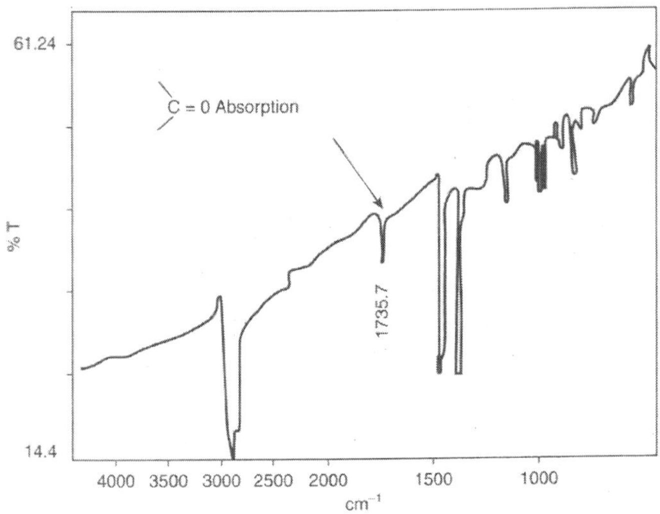

Figure: IR spectrum showing carbonyl absorption due to oxidative degradation of polypropylene crutch moulding

Polymers are susceptible to attack by atmospheric oxygen, especially at elevated temperatures encountered during processing to shape. Many process methods such as extrusion and injection moulding involve pumping molten polymer into tools, and the high temperatures needed for melting may result in oxidation unless precautions are taken. For example, a forearm crutch suddenly snapped and the user was severely injured in the resulting fall. The crutch had fractured across a polypropylene insert within the aluminium tube of the device, and infra-red spectroscopy of the material showed that it had oxidised, possible as a result of poor moulding. Oxidation is usually relatively easy to detect, owing to the strong absorption by the carbonyl group in the spectrum of polyolefins. Polypropylene has a relatively simple spectrum, with few peaks at the carbonyl position (like polyethylene). Oxidation tends to start at tertiary carbon atoms because free radicals here at more stable, so last longer and are attacked by oxygen. The carbonyl group can be further oxidised to break the chain, so weakening the material by lowering the molecular weight, and cracks start to grow in the regions affected.

Ozonolysis

The reaction occurring between double bonds and ozone is known as ozonolysis when one molecule of the gas reacts with the double bond:

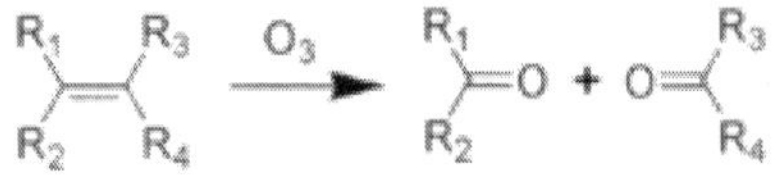

The immediate result is formation of an ozonide, which then decomposes rapidly so that the double bond is cleaved. This is the critical step in chain breakage when polymers are attacked. The strength of polymers depends on the chain molecular weight or degree of polymerization: The higher the chain length the greater the mechanical strength (such as tensile strength). By cleaving the chain, the molecular weight drops rapidly and there comes a point when it has little strength whatsoever, and a crack forms. Further attack occurs in the freshly exposed crack surfaces and the crack grows steadily until it completes a circuit and the product separates or fails. In the case of a seal or a tube, failure occurs when the wall of the device is penetrated.

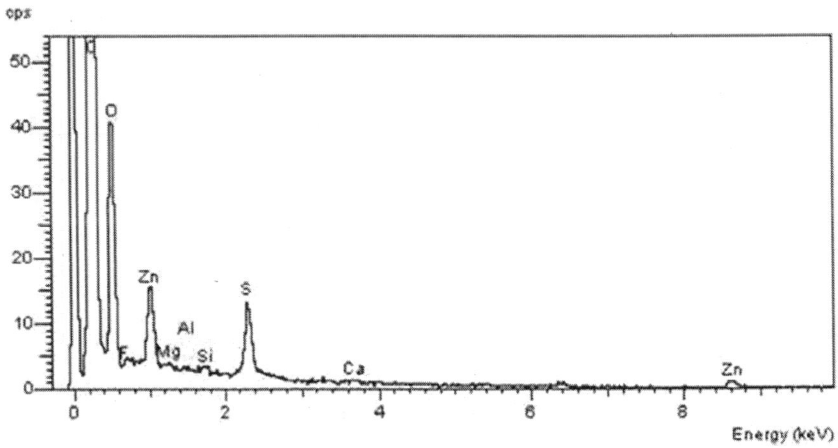

Figure: EDX spectrum of crack surface

EDX Spectrum of Unaffected Rubber Surface: The carbonyl end groups that are formed are usually aldehydes or ketones, which can oxidise further to carboxylic acids. The net result is a high concentration of elemental oxygen on the crack surfaces, which can be detected using Energy-dispersive X-ray spectroscopy in the environmental SEM, or ESEM. The spectrum at left shows the high-oxygen peak compared with a constant sulphur peak. The spectrum at right shows the unaffected elastomer surface spectrum, with a relatively low-oxygen peak compared with the sulphur peak. The spectra were obtained during an investigation into ozone cracking of diaphragm seals in a semi-conductor fabrication factory.

Spectroscopic Notation

Spectroscopic notation provides various ways to specify atomic ionization states, as well as atomic and molecular orbitals.

Atomic and Molecular Orbitals

letter	*name*	*l*
s	sharp	0
p	principal	1
d	diffuse	2
f	fundamental	3
g		4
h		5
i		6
k		7
l		8
m		9
n		10
o		11
q		12
r		13
t		14
u		15
v		16
...		...

Before atomic orbitals were understood, spectroscopists discovered various distinctive series of spectral lines in atomic spectra, which they identified by letters. These letters were later associated with the azimuthal quantum number, *l*. The letters, "s", "p", "≤, and "f", for the first four values of *l* were chosen to be the first letters of properties of the spectral series observed in alkali metals. Letters for subsequent values of l were assigned in alphabetical order, omitting the letter "j":

This notation is used to specify electron configurations and to create the term symbol for the electron states in a multi-electron atom. When writing a term symbol, the above scheme for a single electron's orbital quantum number is applied to the total orbital angular momentum associated to an electron state.

Ionization States

Spectroscopists customarily refer to the spectrum arising from a given ionization state of a given element by the element's symbol followed by a Roman numeral. The numeral I is used for spectral lines associated with the neutral element, II for those from the first ionization state, III for those from the second ionization state, and so on. For example, 'He I' denotes lines of neutral helium, and 'C IV' denotes lines arising from the third ionization state, C^{+++}, of carbon.

Mnemonic

One mnemonic phrase for remembering the sequence *S. P. D. F. G. H. ...* is "Sober Physicists Don't Find Giraffes Hiding In Kitchens Like Mine". One is expected to remember that after the *F*, the letters in the sequence are alphabetical (excluding J).

There are a few other mnemonic phrases for remembering the sequence, for example:

- Smart People Don't Fail
- Silly People Drive Fast

Another mnemonic is: if you don't memorize the proper order, you will get a SPeeDy F!

Molecular Spectroscopic Notation

The spectroscopic notation of molecules uses Greek letters to represent the modulus of the orbital angular momentum along the internuclear axis. The quantum number that represents this momentum is the Ω.

Ω=0,1,2,3,... Symbols: Σ,Π,Δ,Φ

For Σ states, one denotes if there is a reflection in a plane containing the nuclei (symmetric), using the + above. The - is used to indicate that there is not.

For homonuclear diatomic molecules, the index g or u denotes the existence of a centre of symmetry (or inversion centre), and indicates the parity of the total wave function. States of even parity with respect to inversion are denoted g for gerade (German for even), and states of odd parity are denoted u for ungerade (German for odd).

In molecular physics, the molecular term symbol is a shorthand expression of the group representation and angular momenta that characterise the state of a molecule, i.e. its electronic quantum state which is an eigenstate of the electronic molecular Hamiltonian. It is the equivalent of the term symbol for the atomic case. However, the following presentation is restricted to the case of homonuclear diatomic molecules, or symmetric molecules with an inversion centre. For heteronuclear diatomic molecules, the u/g symbol does not correspond to any exact symmetry of the electronic molecular Hamiltonian. In the case of less symmetric molecules the molecular term symbol contains the symbol of the group representation to which the molecular electronic state belongs.

It has the general form:

$$^{2S+1}\Lambda^{(+/-)}_{\Omega,(g/u)}$$

where

- S is the total spin quantum number
- Ω is the projection of the orbital angular momentum along the internuclear axis
- Ω is the projection of the total angular momentum along the internuclear axis
- u/g is the parity
- +/– is the reflection symmetry along an arbitrary plane containing the internuclear axis

Ω Quantum Number

For atoms, we use S, L, J and M_J to characterise a given state. In linear molecules, however, the lack of spherical symmetry destroys the relationship $[\hat{\mathbf{L}}^2, \hat{H}] = 0$, so L ceases to be a good quantum number. A new set of operators have to be used instead: $\{\hat{\mathbf{S}}^2, \hat{\mathbf{S}}_z, \hat{\mathbf{L}}_z, \hat{\mathbf{J}}_z = \hat{\mathbf{S}}_z + \hat{\mathbf{L}}_z\}$, where the z-axis is defined along the internuclear axis of the molecule. Since these operators commute with each other and with the Hamiltonian on the limit of negligible spin-orbit coupling, their eigenvalues may be used to describe a molecule state through the quantum numbers S, M_S, M_L and M_J.

The cylindrical symmetry of a linear molecule ensures that positive and negative values of a given m_l for an electron in a molecular orbital will be degenerate in the absence of spin-orbit coupling. Different molecular orbitals are classified with a new quantum number, λ, defined as

$$\lambda = |m_l|$$

Following the spectroscopic notation pattern, molecular orbitals are designated by a smallcase Greek letter: for λ = 0, 1, 2, 3,... orbitals are called σ, π, δ, φ... respectively.

Now, the total z-projection of L can be defined as

$$M_L = \sum_i m_{li}.$$

As states with positive and negative values of M_L are degenerate, we define

$$\Omega = |M_L|,$$

and a capital Greek letter is used to refer to each value: Ω = 0, 1, 2, 3... are coded as Σ, Π, Δ, Φ... respectively. The molecular term symbol is then defined as

$$^{2S+1}\Omega$$

and the number of electron degenerate states (under the absence of spin-orbit coupling) corresponding to this term symbol is given by:

- $(2S+1)\times 2$ if Ω is not 0
- $(2S+1)$ if Ω is 0.

Ω and Spin–Orbit Coupling

Spin–orbit coupling lifts the degeneracy of the electronic states. This is because the z-component of spin interacts with the z-component of the orbital angular momentum, generating a total electronic angular momentum along the molecule axis J_z. This is characterised by the M_J quantum number, where

$$M_J = M_S + M_L.$$

Again, positive and negative values of M_J are degenerate, so the pairs (M_L, M_S) and ($-M_L$, $-M_S$) are degenerate: {(1, 1/2), (−1, −1/2)}, and {(1, −1/2), (−1, 1/2)} represent two different degenerate states. These pairs are grouped together with the quantum number Ω, which is defined as the sum of the pair of values (M_L, M_S) for which M_L is positive. Sometimes the equation

$$\Omega = \Omega + M_S$$

is used (often Σ is used instead of M_S). Note that although this gives correct values for Ω it could be misleading, as obtained values do not correspond to states indicated by a given pair of values (M_L,M_S). For example, a state with (–1, –1/2) would give an Ω value of $\Omega = |-1| + (-1/2) = -1/2$, which is wrong. Choosing the pair of values with M_L positive will give a Ω = 3/2 for that state.

With this, a level is given by

$$^{2S+1}\Lambda_{\Omega}$$

Note that Ω can have negative values and subscripts r and i represent regular (normal) and inverted multiplets, respectively. For a $^4\Pi$ term there are four degenerate (M_L, M_S) pairs: {(1, 3/2), (–1, –3/2)}, {(1, 1/2), (–1, –1/2)}, {(1, –1/2), (–1, 1/2)}, {(1, –3/2), (–1, 3/2)}. These correspond to Ω values of 5/2, 3/2, 1/2 and –1/2, respectively. Approximating the spin–orbit Hamiltonian to first order perturbation theory, the energy level is given by

$$E = A\ M_L\ M_S$$

where A is the spin–orbit constant. For $^4\Pi$ the Ω values 5/2, 3/2, 1/2 and –1/2 correspond to energies of $3A/2$, $A/2$, $-A/2$ and $-3A/2$. Despite of having the same magnitude, levels of $\Omega = \pm 1/2$ have different energies associated, so they are not degenerate. With that convention, states with different energies are given different Ω values. For states with positive values of A (which are said to be *regular*), increasing values of Ω correspond to increasing values of energies; on the other hand, with A negative (said to be *inverted*) the energy order is reversed. Including higher-order effects can lead to a spin-orbital levels or energy that do not even follow the increasing value of Ω.

When Ω = 0 there is no spin–orbit splitting to first order in perturbation theory, as the associated energy is zero. So for a given S, all of its M_S values are degenerate. This degeneracy is lifted when spin–orbit interaction is treated to higher order in perturbation theory, but still states with same $|M_S|$ are degenerate in a non-rotating molecule. We can speak of a $^5\Sigma_2$ substate, a $^5\Sigma_1$ substate or a $^5\Sigma_0$ substate. Except for the case $\Omega = 0$, these substates have a degeneracy of 2.

Reflection Through a Plane Containing the Internuclear Axis

There are an infinite number of planes containing the internuclear axis and hence there are an infinite number of possible reflections. For any of these planes, molecular terms with $\Omega > 0$ always have a state which is symmetric with respect to this reflection and one state that is antisymmetric. Rather than labelling those situations as, e.g., $^2\Pi^{\pm}$, the $\pm$ is omitted.

For the Σ states, however, this two-fold degeneracy disappears, and all Σ states are either symmetric under any plane containing the internuclear axis, or antisymmetric. These two situations are labelled as Σ^+ or Σ^-.

Reflection Through an Inversion Centre: u and g Symmetry

Taking the molecule centre of mass as origin of coordinates, consider the change of all electrons' position from (x_i, y_i, z_i) to $(-x_i, -y_i, -z_i)$. If the resulting wave function is unchanged, it is said to be *gerade* (German for even); if the wave function changes sign then it is said to be *ungerade* (odd). For a molecule with a centre of inversion, all orbitals will be symmetric or antisymmetric. The resulting wavefunction for the whole multielectron system will be *gerade* if an even number of electrons is in *ungerade* orbitals, and *ungerade* if there is an odd number of electrons in *ungerade* orbitals, independently of the number of electrons in *gerade* orbitals.

Alternative Empirical Notation

Electronic states are also often identified by an empirical single-letter label. The ground state is labelled X, excited states of the same multiplicity (i.e., having the same spin quantum number) are labelled in ascending order of energy with capital letters A,B,C...; excited states having different multiplicity than the ground state are labelled with lower-case letters a,b,c... In polyatomic molecules (but not in diatomic) it is customary to add a tilde (e.g. X˜, a˜) to these empirical labels to prevent possible confusion with symmetry labels based on group representations.

Chapter 2

Spectrum Analysis

Spectrum, also known as emission spectrochemical analysis, is the original scientific method of charting and analysing the chemical properties of matter and gases by looking at the bands in their optical spectrum. The empirical laws of spectrum analysis are commonly known as *Kirchhoff's Three Laws of Spectroscopy* as follows:

1. An incandescent solid, liquid, or gas under high pressure emits a continuous spectrum.
2. A hot gas under low pressure emits a "bright-line" or emission-line spectrum.
3. A continuous spectrum source viewed through a cool, low-density gas produces an absorption-line spectrum.

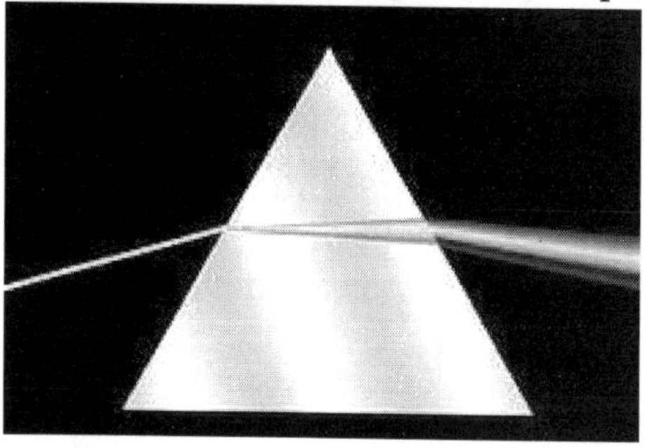

Figure: Light can be separated into a spectrum by a prism. The resulting spectrum can then be analysed

Anders Jonas Ångström, Charles Wheatstone, Gustav Kirchhoff, Robert Bunsen, William Crookes, and others contributed to early spectroscopy through the discovery and exploitation of spectral emission lines. In 1835, Charles Wheatstone reported that different metals could be easily distinguished by the different bright lines in the emission spectra of their sparks, thereby launching the science of spectrum analysis. In 1854, David Alter, a scientist of Freeport,

Pennsylvania, published a work that included the spectral radiance properties for twelve metals, titled *On Certain Physical Properties of Light Produced by the Combustion of Different Metals in an Electric Spark Refracted by a Prism.*

Dr. Alter began studying the optical properties of matter ever since finding a piece of melted, prismatic glass in the debris of the great Pittsburgh fire of 1845. By 1855, Alter published another article that expanded his original theory by including six gases, including the first discovery of what came to be named the Balmer lines of hydrogen. Alter's article contains a paragraph where he visualized the application of spectrum analysis to astronomy, mentioning the study and detection of elements in the combustion of shooting stars or luminous meteors, and daguerreotyped the dark lines of the solar spectrum. Alter's spectral discoveries were noted in various scientific publications in France, Germany, and Switzerland from 1854 to 1860.

In 1853, the Swedish physisist Anders Jonas Ångström presented similar theories about gases having spectra in his work: *Optiska Undersökningar* to the Royal Swedish Academy of Sciences—pointing out that the electric spark yields two superposed spectra. Ångström postulated that an incandescent gas emits luminous rays of the same refrangibility as those it can absorb—a fundamental principle of spectrum analysis.

In 1860, German physicist Gustav Kirchhoff and chemist Robert Bunsen published their own findings on the spectra of eight metals and identified these metals in natural elements. Kirchoff went on to contribute fundamental research on the nature of spectral absorption. Spectrum analysis was then grouped by Kirchhoff into the three fundamental laws commonly called Kirchoff's Laws, these laws integrated both Alter and Ångström's discoveries of radiance and emission with Kirchhoff's fundamental discoveries of absorption.

Johann Balmer discovered in 1885 that the four visible lines of hydrogen were part of a series that could be expressed in terms of integers. This was followed a few years later by the Rydberg formula, which described additional series of lines.

In the 1860's William Huggins and his wife Margaret used spectroscopy to determine that the stars were composed of the same elements as found on earth. They also used the non-relativistic Doppler shift (redshift) equation on the spectrum of the star Sirius in 1868 to determine its axial speed. They were the first to take a spectrum of a planetary nebula when the Cat's Eye Nebula (NGC 6543) was analysed. Using spectral techniques, they were able to distinguish nebulae from galaxies. In the early twentieth century, spectrum analysis led to "atomic spectroscopy" and quantum mechanics.

Spectral Power Distribution

In colour science and radiometry, a spectral power distribution (SPD) describes the power per unit area per unit wavelength of an illumination (radiant exitance), or more generally, the per-wavelength contribution to any radiometric quantity (radiant energy, radiant flux, radiant intensity, radiance, irradiance, radiant exitance, or radiosity).

Mathematically, for the spectral power distribution of a radiant exitance or irradiance one may write:

$$M_\lambda = \frac{\partial^2 \Phi}{\partial A \partial \lambda} \approx \frac{\Phi}{A \Delta \lambda}$$

where $M(\lambda)$ is the spectral irradiance (or exitance) of the light (SI units: W/m^3 = kg/(m·s^3)); Φ is the radiant flux of the source (SI unit: watt, W); A is the area over which the radiant flux is integrated (SI unit: square metre, m^2); and λ is the wavelength (SI unit: metre, m). (Note that it is more convenient to express the wavelength of light in terms of nanometres; spectral exitance would then be expressed in units of W·m^{-2}·nm^{-1}.) The approximation is valid when the area and wavelength interval are small.

Relative SPD

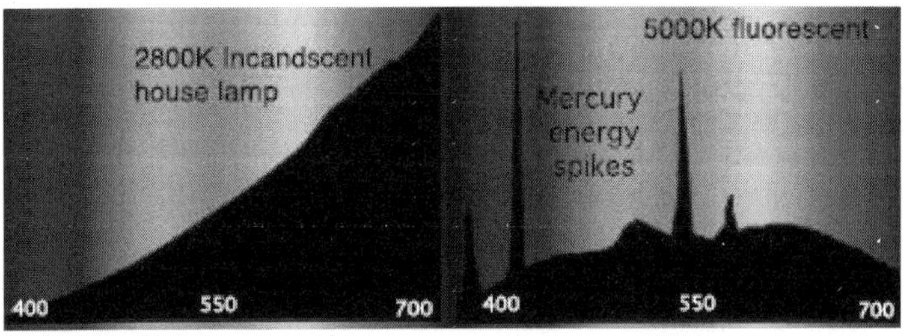

Figure: Characteristic spectral power distributions (SPDs) for an incandescent lamp (left) and a fluorescent lamp (right). The horizontal axes are in nanometres and the vertical axes show relative intensity in arbitrary units.

Because the luminance of lighting fixtures and other light sources are handled separately, a spectral power distribution may be normalized in some manner, often to unity at 555 or 560 nanometres, coinciding with the peak of the eye's luminosity function.

Scattering Theory

Top: the real part of a plane wave travelling upwards. Bottom: The real part of the field after inserting in the path of the plane wave a small transparent disk of index of refraction higher than the index of the surrounding medium. This object scatters part of the wave field, although at any individual point, the wave's frequency and wavelength remain intact.

In mathematics and physics, scattering theory is a framework for studying and understanding the scattering of waves and particles. Prosaically, wave scattering corresponds to the collision and scattering of a wave with some material object, for instance sunlight scattered by rain drops to form a rainbow.

Scattering also includes the interaction of billiard balls on a table, the Rutherford scattering (or angle change) of alpha particles by gold nuclei, the Bragg scattering (or diffraction) of electrons and X-rays by a cluster of atoms, and the inelastic scattering of a fission fragment as it traverses a thin foil.

More precisely, scattering consists of the study of how solutions of partial differential equations, propagating freely "in the distant past", come together and interact with one another or with a boundary condition, and then propagate away "to the distant future".

The direct scattering problem is the problem determining the distribution of scattered radiation/particle flux basing on the characteristics of the scatterer.

The inverse scattering problem is the problem of determining the characteristics of an object (e.g., its shape, internal constitution) from measurement data of radiation or particles scattered from the object. Since its early statement for radiolocation, the problem has found vast number of applications, such as echolocation, geophysical survey, nondestructive testing, medical imaging and quantum field theory, to name just a few.

Conceptual Underpinnings

The concepts used in scattering theory go by different names in different fields. The object of this section is to point the reader to common threads.

Composite Targets and Range Equations

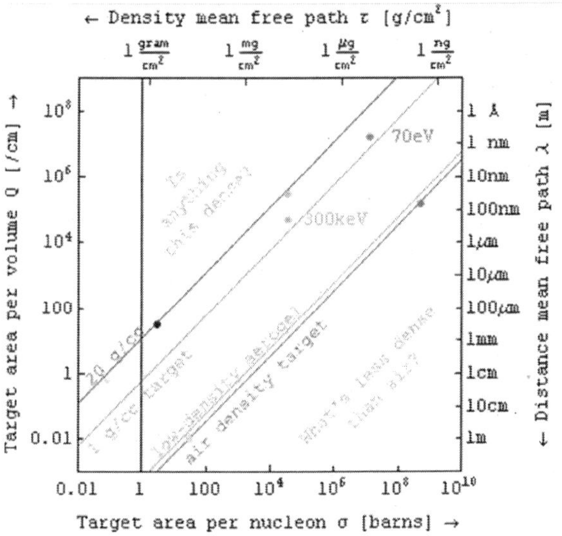

Figure: Equivalent quantities used in the theory of scattering from composite specimens, but with a variety of units.

When the target is a set of many scattering centres whose relative position varies unpredictably, it is customary to think of a range equation whose arguments take different forms in different application areas. In the simplest case consider an interaction that removes particles from the "unscattered beam" at a uniform rate that is proportional to the incident flux *I* of particles per unit area per unit time, i.e. that

$$\frac{dI}{dx} = -QI$$

where Q is an interaction coefficient and x is the distance travelled in the target. The above ordinary first-order differential equation has solutions of the form:

$$I = I_o e^{-Q\Delta x} = I_o e^{-\frac{\Delta x}{\lambda}} = I_o e^{-\sigma(\eta\Delta x)} = I_o e^{-\frac{\rho\Delta x}{\tau}},$$

where I_0 is the initial flux, path length Δxa"x-x_0, the second equality defines an interaction mean free path λ, the third uses the number of targets per unit volume η to define an area cross-section σ, and the last uses the target mass density ρ to define a density mean free path τ. Hence one converts between these quantities via Q = 1/ λ = ησ = ρ/τ, as shown in the figure at left.

In electromagnetic absorption spectroscopy, for example, interaction coefficient (e.g. Q in cm^{-1}) is variously called opacity, absorption coefficient, and attenuation coefficient. In nuclear physics, area cross-sections (e.g. σ in barns or units of 10^{-24} cm^2), density mean free path (e.g. τ in grams/cm^2), and its reciprocal the mass attenuation coefficient (e.g. in cm^2/gram) or *area per nucleon* are all popular, while in electron microscopy the inelastic mean free path (e.g. λ in nanometres) is often discussed instead.

In Theoretical Physics

In mathematical physics, scattering theory is a framework for studying and understanding the interaction or scattering of solutions to partial differential equations. In acoustics, the differential equation is the wave equation, and scattering studies how its solutions, the sound waves, scatter from solid objects or propagate through non-uniform media (such as sound waves, in sea water, coming from a submarine). In the case of classical electrodynamics, the differential equation is again the wave equation, and the scattering of light or radio waves is studied. In quantum mechanics and particle physics, the equations are those of quantum electrodynamics QED, quantum chromodynamics QCD and the Standard Model, the solutions of which correspond to fundamental particles. In quantum chemistry, the solutions correspond to atoms and molecules, governed by the Schrödinger equation.

Elastic and Inelastic Scattering: The example of scattering in quantum chemistry is particularly instructive, as the theory is reasonably complex while still having a good foundation on which to build an intuitive understanding. When two atoms are scattered off one another, one can understand them as being the bound state solutions of some differential equation. Thus, for example, the hydrogen atom corresponds to a solution to the Schrödinger equation with a negative inverse-power (i.e., attractive Coulombic) central potential. The scattering of two hydrogen atoms will disturb the state of each atom, resulting in one or both becoming excited, or even ionized. Thus, collisions can be either elastic (the internal quantum states of the particles are not changed) or inelastic (the internal quantum states of the particles are changed). From the experimental viewpoint the observable quantity is the cross section. From the theoretical viewpoint the key quantity is the S matrix.

According to the optics classification of the Optical Society of America this field consists of the following topics:

- Aerosol and cloud effects
- Atmospheric scattering
- Backscattering
- Diffusion
- Extinction
- Index measurements
- Inverse scattering
- Linewidth
- Long-wave scattering
- Mie theory
- Multiple scattering
- Scattering measurements
- Brillouin scattering
- Molecular scattering
- Particle scattering
- Raman scattering
- Rayleigh Scattering
- Scattering from rough surfaces
- Stimulated scattering
- Stimulated Brillouin scattering
- Stimulated Raman scattering
- Scintillation
- Turbid media.

The Mathematical Framework: In mathematics, scattering theory deals with a more abstract formulation of the same set of concepts. For example, if a differential equation is known to have some simple, localised solutions, and the solutions are a function of a single parameter, that parameter can take the conceptual role of time. One then asks what might happen if two such solutions are set up far away from each other, in the "distant past", and are made to move towards each other, interact (under the constraint of the differential equation) and then move apart in the "future". The scattering matrix then pairs solutions in the "distant past" to those in the "distant future".

Solutions to differential equations are often posed on manifolds. Frequently, the means to the solution requires the study of the spectrum of an operator on the manifold. As a result, the solutions often have a spectrum that can be identified with a Hilbert space, and scattering is described by a certain map, the S matrix, on Hilbert spaces. Spaces with a discrete spectrum correspond to bound states in quantum mechanics, while a continuous spectrum is associated with scattering states. The study of inelastic scattering then asks how discrete and continuous spectra are mixed together. An important, notable development is the inverse scattering transform, central to the solution of many exactly solvable models.

Element Identification and Emission Spectra

The energy levels in atoms and ions are the key to the production and detection of light. Energy levels or "shells"exist for electrons in atoms and molecules. The colours of dyes and other compounds results from electron jumps between these shells or levels. The colours of fireworks result from jumps of electrons from one shell to another. Observations of light emitted by the elements are also evidence for the existence of shells, subshsells and energy levels. The kinds of light that interact with atoms indicate the energy differences between shells and energy levels in the quantum theory model of the atom. Typically the valence electrons are the ones involved in these jumps.

The "quantum" theory was proposed more than 90 years ago, and has been confirmed by thousands of experiments. Science and education has failed to clearly describe the energy level concept to almost four generations of citizens. This experiment is an exercise aimed at throwing a little more light on the subject. Atoms have two kinds of states; a ground state and an excited state. The ground state is the state in which the electrons in the atom are in their lowest energy levels possible (atoms naturally are in the ground state). This means the electrons have the lowest possible values for "n" the principal quantum number.

Specific quantized amounts of energy are needed to excite an electron in an atom and produce an excited state. The animation shows the opposite of excitation. It shows how the excited hydrogen atom with an electron in the n = 3 shell can release energy. If the electron in hydrogen only drops to the n = 2 shell the energy matches a pulse of red light. Note the size of the electron cloud in the excited atom changes when the electron moves from shell to shell. The size of the atom decreases in volume when the electron goes from the n=3 shell to the n = 2 shell. On average the electrons are closer to the nucleus for lower values of "n".

The electron cloud is related to the most probable distance between the nucleus and the electron. The most probable distance increases with increasing "n" value. The excited electron is still "in" the atom even in an excited state. The valence electron will only escape the atom if the electron is given an amount of energy equal to the ionization energy for that atom.

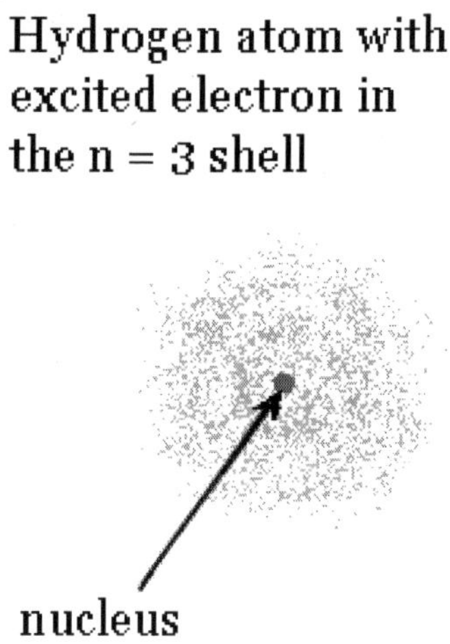

Figure:

Energy can be added to atoms many different ways. It can be in the form of light, an electric discharge or heat. This added or extra energy is emitted when the excited electrons in the atoms give off light and fall back to lower shells. The light emitted has wavelengths and colours that depend on the amount of energy originally absorbed by the atoms. Usually each individual excited atom will emit one type of light. Since we have billions and billions of atoms we get billions of excitations and emissions.

Not all atoms in a sample will absorb or be excited exactly the same. For example in hydrogen the ground state has the electron in the n= 1 shell. The electron in some hydrogen atoms may be excited into the n = 2 level. Other hydrogen atoms can have the electron excited into the n = 4 shell. Different elements emit different emission spectra when they are excited because each type of element has a unique energy shell or energy level system. Each element has a different set of emission colours because they have different energy level spacings. We will see the emission spectra or pattern of wavelengths (atomic spectra) emitted by six different elements in this lab. We will then identify an unknown element by comparing the colour of the unknown with the flame colour of our knowns. You need to know that white light is the combination of all colours of the spectrum.

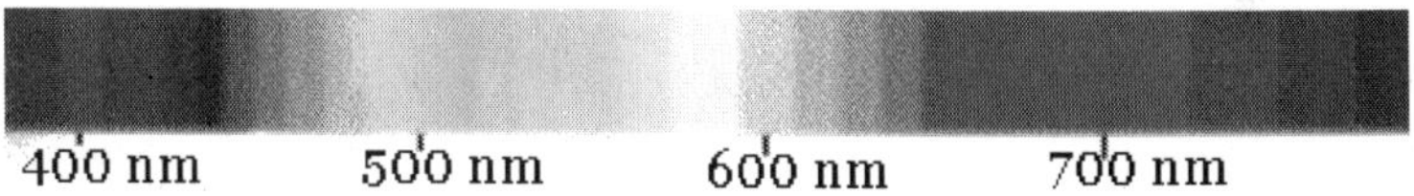

Figure:

Each colour has a characteristic wavelength. The wavelength is the distance between the beginning and end of a complete cycle of the light wave. All colours of light travel at the same speed, 3.0 x 10^8 metres/second. The animation shows how a prism separates photons of red light from photons of blue light. The photons of different colours fall in different positions on the colour spectrum. The position is determined by the wavelength. (Be sure to view Animation)

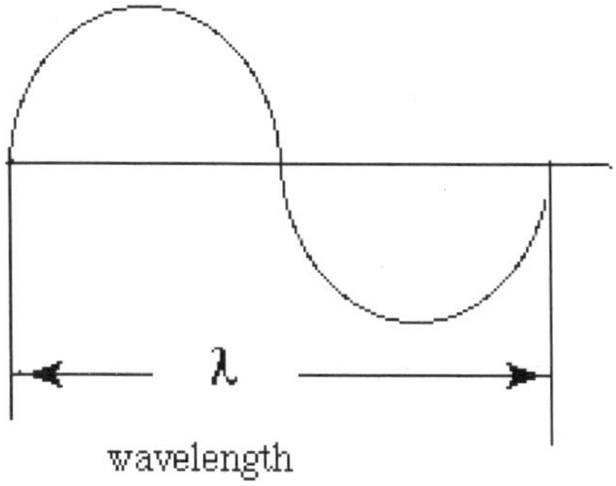

Figure: The wavelength for a particular colour of visible light is redc off the calibration on the scale for the spectrum.

The position the scale indicates λ the wavelength.

The speed of light frequency and wavelength are related by the formula

c = l n

where c is the speed of light 3.00 x 10^8 metres/second.

n is the frequency of the light wave in cycles per second

Lambda, l, is the wavelength in metres/cycle.

The frequency,n, equals the number of cycles that pass an observer in a second.

These individual cycles each carry a specific small amount of energy. The particles of light are called photons.

The energy of a light photon or particle is different for each colour. If you have ever seen a Star Trek movie you may have heard of "photon torpedoes". The science fiction writers suggest that bundles of light energy could be harnessed and used as projectiles.

We can calculate the energy of a photon or wave packet using the Planck energy relation shown here. Planck won a Nobel Prize for this work. This relation allows us to tell how much energy a single photon can carry.

$E = h\ n$

here E is the energy of the photon in joules

h is Planck's constant 6.624×10^{-34} joule seconds

n is the frequency of the particular light wave

Red light has longer wavelength and is lower in energy than blue light. The wavelength of red light corresponds to the range of 700 to 600 nanometres, (7000 Ångstrom or 0.0000007 metres).

Blue light has shorter wavelength in the range of 400 nm (4000 Ångstrom or 0.00000004 metre,

$1\ Å = 1 \times 10^{-10}$ m = 0.0000000001 metre = 1×10^{-1} nanometre).

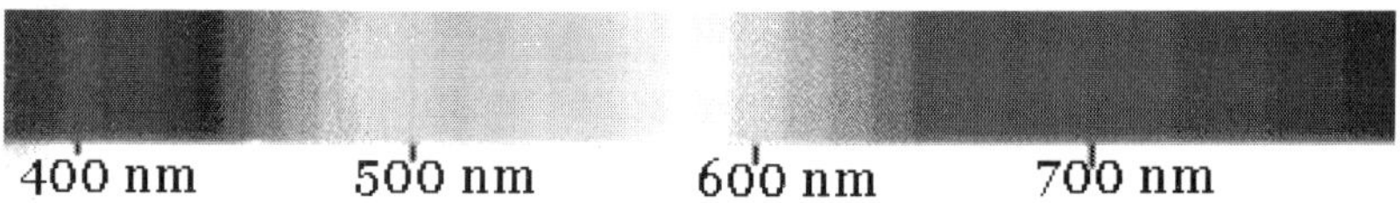

Figure:

Spectroscopy is the analysis of light spectra and the way in which light interacts with matter.

When light is analysed it is commonly separated into its component colours. The light source is directed on a slit and the "beam" of light is separated using a prism or grating. The reason that the images are lines is that the light from the lamp is focused on a narrow slit. The illustration shows the separation of a light beam into its component colours.

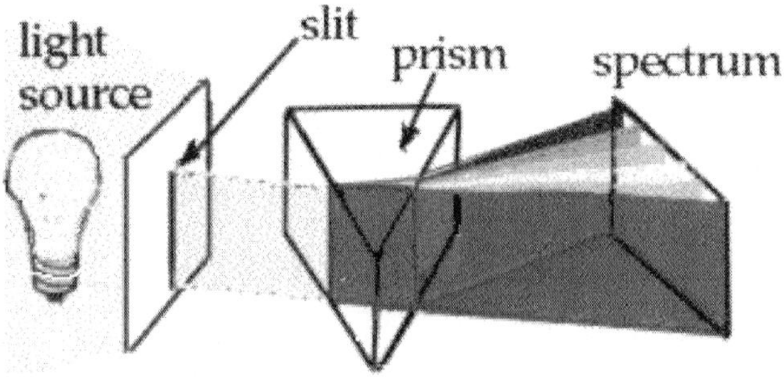

Figure:

This produces an image of the slit which has the shape of a line. The resulting beam of light can be broken into the colour spectrum or into its components of the spectrum emitted by the atom. You can see the specific colours emitted by the light source. A white light source will give a spectrum like the one shown above. The emission spectrum for sodium shows only two colours in the visible colour range. The two colours are yellow and have wavelengths of about 590 nm. The continuous spectrum and calibration scale is shown to give approximate wavelength values. A CD-ROM mirrored surface behaves like a grating or prism. If you look at the surface of the CD-ROM under a light, you can see the colour spectrum. Street lights are sodium vapour lamps in many communities. These lamps have an orange yellow tint. You can see from the emission spectrum why the sodium vapour lamps would appear yellow and not white. These lamps consume less energy than the older blue coloured mercury vapour lamps. Mercury vapour lamps have been sold in hardware stores for yard lighting.

One of the odd things that we sometimes notice is that coloured things have a different appearance in natural daylight than they do under mercury vapour or sodium vapour lamps. This is reasonable because the daylight includes all of the wavelengths of white light and the vapour lights only emit a few specific colours that can be reflected into our eye off of any illuminated article.

Example Calculations

Frequency

How to calculate frequency, n, from wavelength and the speed of light, c. A wave with the colour red has a wavelength of 6.50×10^{-7} metres/cycle), what is the frequency of the wave?

Use c = l n solve for n and we get the following n = {c/l }

n = {c/l } = (3.00×10^{8} metres/second)/(6.50×10^{-7} metres/cycle)

n = 4.62×10^{14} cycles/second with three sf

The energy of a photon or single pulse of light energy can be calculated using the Planck energy relation

E = h n = (6.624×10^{-34} joules/cycle) (4.62×10^{14} cycles/second)

E = 3.05×10^{-19} joules/cycle

Note: A common household 100 watt light bulb uses 100 joules every second. Even though this energy is not converted completely into light, we can make a rough assumption that it is. If this energy is emitted as light (really much is disipated as heat) it means 3.28×10^{20} photons (330,000,000,000,000,000,000 photons) would have to be emitted every second to transfer the 100 joules in a second. The numbers are huge. By way of comparison the population of the world is only 8,000,000,000 (eight billion). The atoms and their electrons are constantly changing energy. The electrons are going through ceaseless energy level jumps.

The Facts of Light

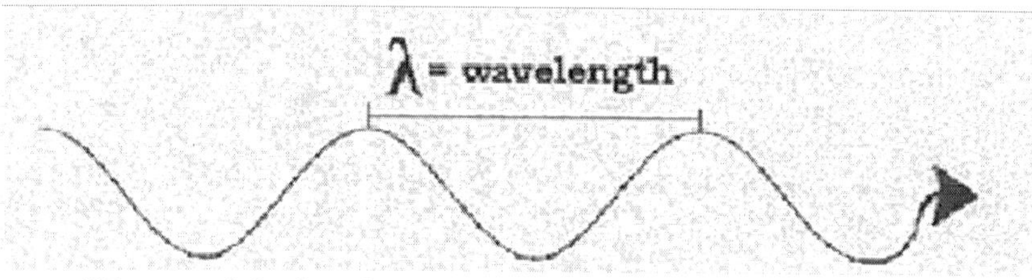

Figure: Wave model of a photon (credit Patricia Knezek)

Light is what enters our eyes and lets us see the world around us. We normally describe light as what we see coming from the sun or out of a light bulb. But what is light? Light is energy- visible light is energy we can see. Scientists have two different models to describe how light behaves. One is the particle model. In this model, we can think of light as a stream of many tiny energy packets called photons (from the Greek word photos, meaning light). Each photon represents a specific amount of energy. The other way to describe light is in the form of a wave. In the wave model, each colour of visible light corresponds to a different wavelength, which is the spacing between two consecutive crests on a wave. The shorter the spacing between the crests, the smaller the wavelength.

These two models of light are related in the following way: the amount of energy a photon contains depends on its wavelength. The smaller the wavelength of the photon, the greater the energy it contains.

When we see visible light, it appears white. This is because we are seeing many different wavelengths of light all at once. If we separate white light by using a prism, we see a rainbow of colours. The colours of a rainbow are always in the same order. We normally divide the rainbow into seven main colours: red, orange, yellow, green, blue, indigo and violet (you can remember their order with the acronym ROY G. BIV). But a rainbow is really a continuous range of colours, since the Sun emits light at all different wavelengths. Each colour corresponds to a specific wavelength of light.

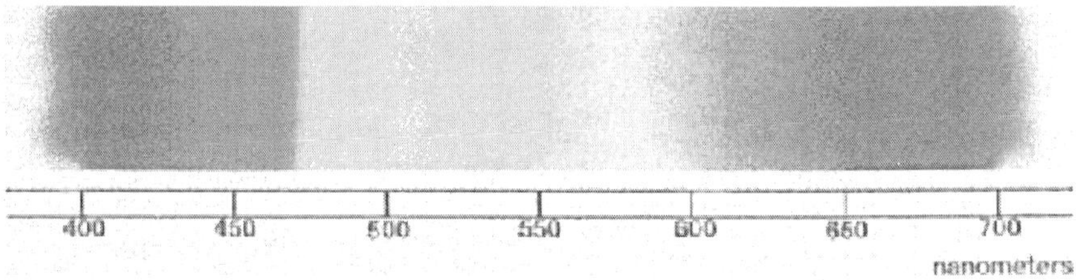

Figure: The Visible Spectrum

The wavelengths of visible light are usually measured in nanometres (nm), or in Angstroms (Å). A nanometre is one-billionth of a metre! And 1 Angstrom is one tenth of a nanometre. Our eyes can see wavelengths of light that range from 400 nm for violet light up to 700 nm for red light. Photons of violet light have more energy than photons of red light, because violet light has a shorter wavelength.

The light that we see with our eyes is actually only one type of light- visible light. There are other forms of light that we cannot see, yet we know they exist. These forms include gamma rays, X-rays, ultraviolet, infrared, microwaves and radio waves. Scientists refer to all the different forms of light as electromagnetic radiation. While we experience these types of radiation in vastly different ways, they are really all the same thing- light! Scientists arrange the kinds of light by the size of their wavelengths in what is called the electromagnetic spectrum.

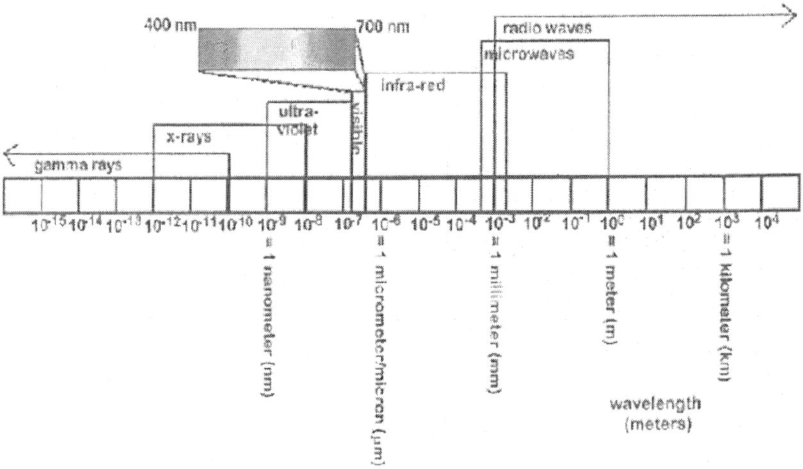

Figure: The Electromagnetic Spectrum

Gamma rays have the shortest wavelengths, about the size of the nucleus of an atom! Gamma ray photons carry a tremendous amount of energy (about ten billion times as much as a photon of visible light!) and therefore can be very harmful to living organisms. X-ray light also has very short wavelengths and very high energy. X-rays are able to pass directly through many materials. While this lets us use X-rays to take images of the bones inside our bodies, it also makes X-ray light harmful to our body cells. Ultraviolet light (which means "beyond violet") is responsible for giving us suntans and sunburns.

The fact that ultraviolet light is able to "burn" our skin reminds us that this type of light has a high level of energy. Too much ultraviolet light is dangerous to most living organisms. Fortunately most ultraviolet light emitted by the sun never reaches us. It is blocked primarily by the ozone layer of the Earth's atmosphere. Glass is also able to block, or absorb, ultraviolet light- which is why no one sunbathes indoors! Infrared, microwaves and radio waves are kinds of light that have lower amounts of energy, because the photon wavelengths get longer and longer.

Even still, we are able to use them in many ways. We experience infrared light (which means "below red") as heat- another form of energy! We sometimes see heat lamps in restaurants used to keep food warm: they emit infrared light. Rescue workers use infrared detectors after earthquakes to locate people who are trapped inside collapsed buildings. Microwaves are used to heat food and to send communication signals for TV and telephone without using wires. Satellite dish antennas are devices that send or receive microwaves. Radio waves have the least energy of all, yet we can send them over long distances to transmit voice and music signals. AM radios receive light waves that are longer than a football field!

The Speed of Light

One of the remarkable things about light is that it always travels at a constant speed in a vacuum (space). Light moves incredibly fast, nearly 300,000 kilometres per second! (That equals 186,000 miles per second!) Because the Sun is 150,000,000 kilometres (93 million miles) away from Earth, light from the Sun takes over eight minutes to reach our eyes. When we see the Sun in the sky, we are actually seeing how the Sun looked eight minutes ago. So the next time you watch the Sun set, keep in mind that it has actually been below the horizon for eight minutes! When we look at other stars and galaxies, it is like looking at a very old photograph. We are seeing them as they looked when they originally emitted the light.

The age of this "photograph" depends on how long the light has had to travel through space to reach our eyes. Astronomers often describe the distance from stars and galaxies to Earth by calculating the time it took their light to reach us. Astronomers use a unit called a light-year. This is the distance that light travels during one year's time. A light-year is equal to about 9,500,000,000,000 kilometres (or 5.9 trillion miles)! Go to a place far away from bright city lights and look at the sky on a clear night. You will see the light from thousands of stars, even without a telescope. The nearest of these stars (other than the Sun, of course) is Alpha Centauri, about 4.3 light-years away from Earth. This means that the distance to this star is about 25 trillion miles. If we launched a spacecraft in the direction of Alpha Centauri, it would take the spacecraft over 25,000 years to reach the star! Since we can't yet send any spacecraft to nearby stars in our lifetime, the only way astronomers can study stars, or even more distant objects like other galaxies, is to study the light that these objects send us from across the universe.

How Astronomers Study Light

When most people think about astronomy, they think about the pictures we see of celestial objects like the Moon and planets, comets, stars and galaxies. Many of these images are formed by the visible light that they give off. Yet these objects also emit light at all other wavelengths, including forms of invisible light such as X-ray, ultraviolet and infrared. In the same way we make photographs using visible light, we can also create images made from the invisible light that objects emit, often using electronic detectors. Astronomers can study objects that emit very little visible light- such as pulsars, quasars and black holes- by using telescopes that "see" the forms of light our eyes cannot. Astronomers are able to uncover information about many of the objects in the universe by comparing images taken in different regions of the electromagnetic spectrum.

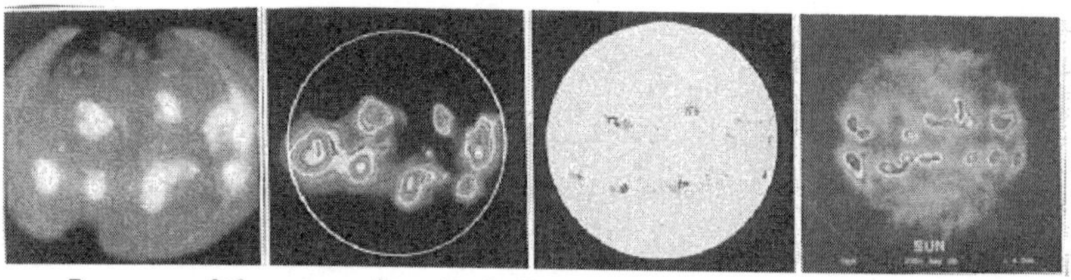

Figure: Images of the sun taken in different regions of the em spectrum; (from left to right) x rays, ultraviolet, visible, and radio

While people are often fascinated by pictures from outer space, the images alone do not always give scientists all the information they need about the object in the picture. In order to study the information contained in different forms of light, astronomers need to further separate light into individual wavelengths. This technique is known as spectroscopy. To separate different wavelengths of light, scientists use prisms or diffraction gratings. Diffraction gratings work just like prisms, but are able to separate wavelengths of light into much finer detail. This allows scientists to study very specific parts of the electromagnetic spectrum one at a time. Scientists also need to measure the amount of light- called the intensity- coming from a certain location. We normally think of intensity as "brightness" of an object. To measure intensity, scientists use electronic devices that count the number of photons striking them. By collecting data on how much light is emitted by an object at specific wavelengths, scientists can obtain information about the properties of the object itself.

Scientists seek to answer many questions about physical objects by studying the amount of light they emit or absorb at different wavelengths. To do this, scientists must establish some relationship between various wavelengths of light and the light intensity they observe at these wavelengths. Relationships are seen most clearly on graphs, so scientists transfer the information of light spectra gathered by electronic detectors onto graphs. By doing this, they can compare numbers directly. If you look at a rainbow spectrum, the intensity of the different colours you see represent the amount of light at each wavelength. The brightness of each colour is related to the intensity at that particular wavelength. In the example below, the intensity of the colours in the rainbow produced by sunlight is graphed according to wavelength. The graph shows that the light is most intense, or brightest, at a wavelength of about 550 nanometres, which is in the yellow region. This is what should be expected, since a rainbow is produced by sunlight, and the Sun appears to be yellow! The reason the Sun appears to be yellow is because it is at that colour that the Sun emits the highest intensity.

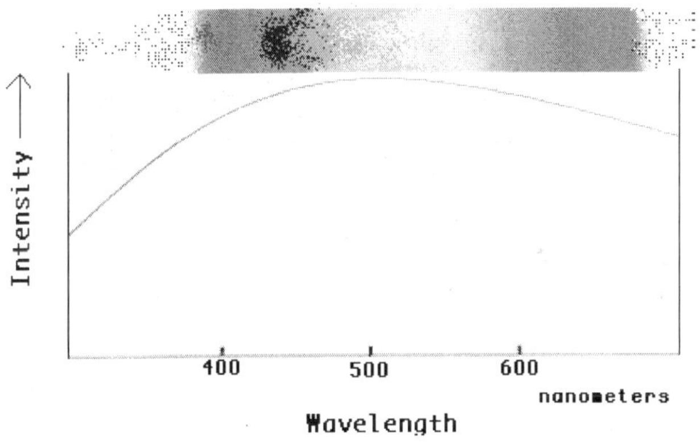

Figure: Continuum Spectrum

Other sources of light, such as gaseous nebulae, do not emit a continuous spectrum with all the colours of the rainbow; they emit light only at discrete wavelengths. These wavelengths of light are like fingerprints. They help us identify what type of substance is producing the light. Think about the neon signs we see in windows- does neon light look like sunlight? If we were to look at neon light through a diffraction grating, we would not see all the colours in the rainbow. Instead, we would only see a few strong lines of colour. The graph below shows the light intensity of a neon sign is high only at a few specific wavelengths. Whenever scientists see this type of pattern on a graph, they know they are looking at neon. This pattern is called the emission spectrum of neon. The spikes or peaks on the graph are called emission lines. What is the relationship between the colour of a neon sign and the wavelengths of light that neon emits?

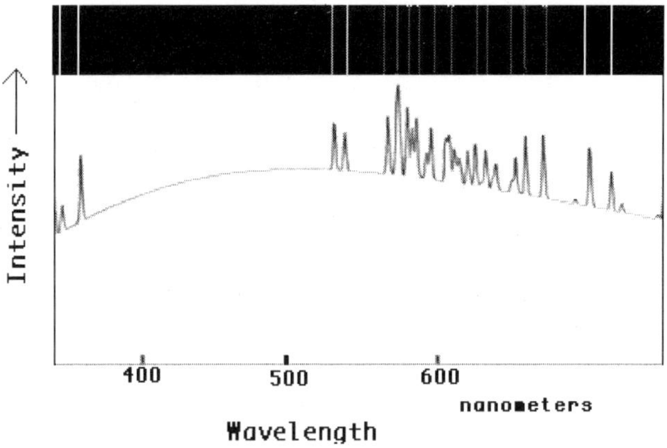

Figure: Neon line emission spectrum

Sometimes chemical elements can be identified by the types of light they absorb. Neon gas emits light in a neon sign because of the electricity passing through it. When the electric current is turned off, the neon sign stops glowing. Now think about sunlight passing through a neon sign that has been turned off. If we were to look at this sunlight through a diffraction grating, we would no longer see a complete rainbow- some of the colours would be missing! This is because the neon gas in the tubes actually absorbed some of the sunlight. If you look at the graph below, you see that the light intensity dips far down at certain wavelengths. This type of pattern is called an absorption spectrum. The dips on the graph are called absorption lines. What do you notice when you compare the absorption and emission lines for neon?

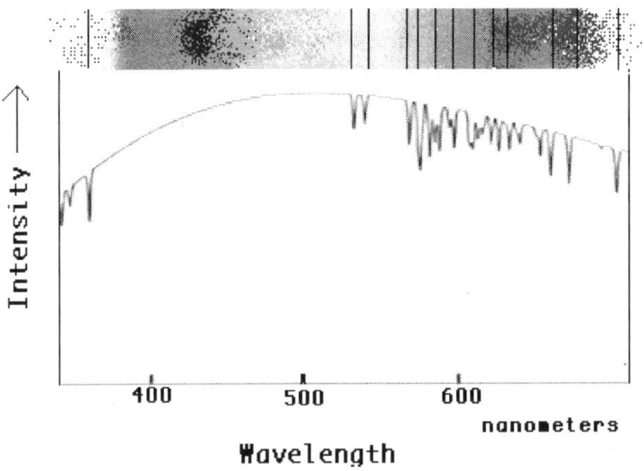

Figure: Neon line absorption spectrum

Every element in the periodic table, and every chemical compound, has its own unique pattern of spectral lines. This pattern is the "fingerprint" scientists use to identify the presence of that given chemical element. Spectroscopy, or the study of the light spectrum, can be used by scientists in many fields. Biologists use spectroscopy to study the pigments in plants; this helps them understand how photosynthesis works. Chemists can look at spectra to identify the types of atoms or molecules present in an unknown substance. And astronomers use spectra to learn about distant objects in the universe. Spectroscopy is often the sole method astronomers can use to do their work. Remember that astronomers can't bring the planets and stars they wish to study into their laboratories! The only way they can learn about the objects in the universe is to study the light they emit. Astronomers use spectroscopy to study properties of objects in the universe, besides chemical composition, such as the temperature, density, chemical composition and speed of a celestial object, and whether it is moving away from or towards us.

The Far Ultraviolet Wavelength Range

To help advance our understanding of the universe, NASA will launch a new space telescope in 1999 called FUSE- the Far Ultraviolet Spectroscopic Explorer. It will examine a part of the electromagnetic spectrum that has not yet been looked at extensively- the far ultraviolet region, where the wavelengths of light range from 90nm to 120nm. The FUSE satellite will orbit outside most of the Earth's atmosphere, just like the Hubble Space Telescope. This space telescope will be a valuable new tool for astronomers, because FUSE is designed to study wavelengths of light too short for Hubble to observe. One purpose of putting telescopes into space is to observe forms of light that do not reach the Earth's surface. (It's a good thing that most electromagnetic radiation is blocked by the Earth's atmosphere; life could not exist if we were constantly bombarded by gamma rays, X-rays and UV light!) The instruments on FUSE will use spectroscopic techniques to study light that does not penetrate Earth's atmosphere. Astronomers are very eager to observe the light in the far ultraviolet region. They will be able to observe many interesting spectral lines in this region. The data they collect will yield clues to how the universe was formed, and how it continues to evolve. People continue to be fascinated by what scientists have learned so far about the universe. Yet many questions remain unanswered. It is through our modern understanding of the behaviour of light that we have learned so much already. Spectroscopy is an extremely important tool that astronomers and other scientists use to study light, and will help us explore the remaining mysteries of the universe.

Atomic Absorption, Emission and Fluorescence Techniques: Atomic spectroscopy is the determination of elemental composition by its electromagnetic or mass spectrum. The study of the electromagnetic spectrum of elements is called Optical Atomic Spectroscopy. Electrons exist in energy levels within an atom. These levels have well defined energies and electrons moving between them must absorb or emit energy equal to the difference between them. In optical spectroscopy, the energy absorbed to move an electron to a more energetic level and/or the energy emitted as the electron moves to a less energetic energy level is in the form of a photon. The wavelength of the emitted radiant energy is directly related to the electronic transition which has occurred. Since every element has a unique electronic structure, the wavelength of light emitted is a unique property of each individual element. As the orbital configuration of a large atom may be complex, there are many electronic transitions which can occur, each transition resulting in the emission of a characteristic wavelength of light, as illustrated below.

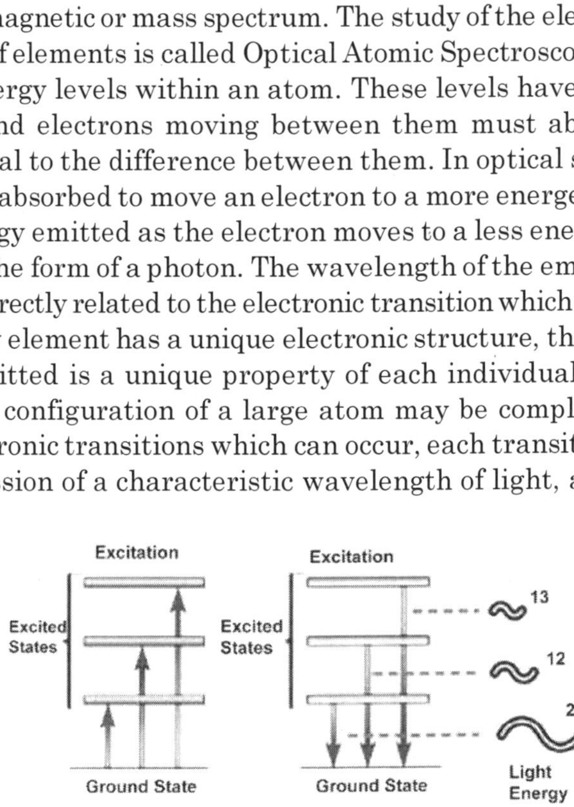

Figure: Energy Transitions

The science of atomic spectroscopy has yielded three techniques for analytical use:

- Atomic Absorption.
- Atomic Emission.
- Atomic Fluorescence.

The process of excitation and decay to the ground state is involved in all three fields of atomic spectroscopy. Either the energy absorbed in the excitation process, or the energy emitted in the decay process is measured and used for analytical purposes.

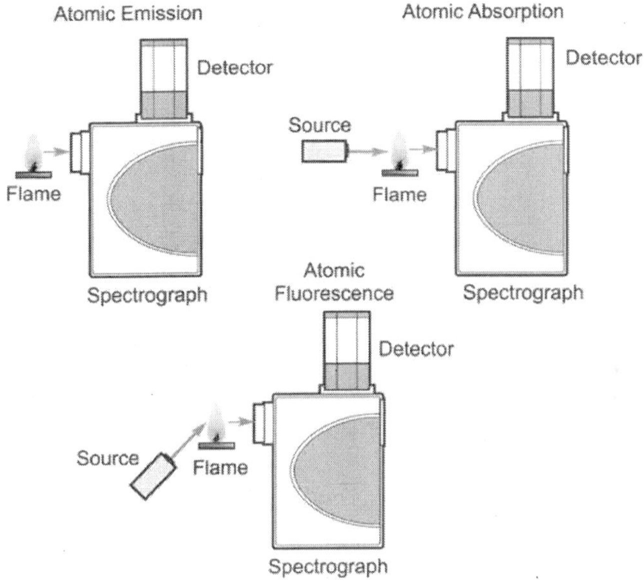

Figure: How the three techniques are implemented.

Atomic Absorption Spectroscopy

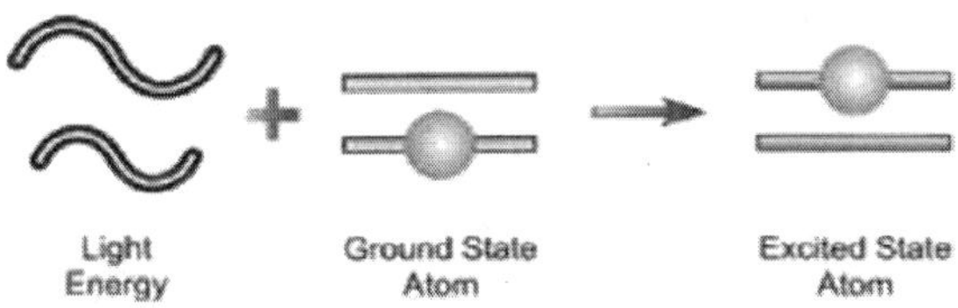

Figure: The atomic absorption process

If light of just the right wavelength impinges on a free, ground state atom, the atom may absorb the light as it enters an excited state in a process known as atomic absorption. This process is illustrated on the right. Atomic absorption measures the amount of light at the resonant wavelength which is absorbed as it passes through a cloud of atoms. As the number of atoms in the light path increases, the amount of light absorbed increases in a predictable way. By measuring the amount of light absorbed, a quantitative determination of the amount of analyte element present can be made. The use of special light sources and careful selection of wavelength allow the specific quantitative determination of individual elements in the presence of others. The atom cloud required for atomic absorption measurements is produced by supplying enough thermal energy to the sample to dissociate the chemical compounds into free atoms. Aspirating a solution of the sample into a flame aligned in the light beam serves this purpose. Under the proper flame conditions, most of the atoms will remain in the ground state form and are capable of absorbing light at the analytical wavelength from a source lamp. The ease and speed at which precise and accurate determinations can be made with this technique have made atomic absorption one of the most popular methods for the determination of metals.

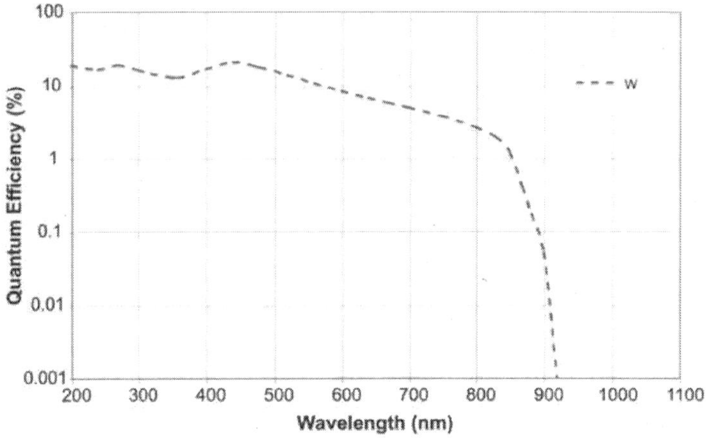

Figure: ICCD Quantum Efficiency relevant to Atomic spectroscopy

Atomic Emission Spectroscopy

In atomic emission, a sample is subjected to a high energy, thermal environment in order to produce excited state atoms, capable of emitting light. The energy source can be an electrical arc, a flame, or more recently, a plasma.

The emission spectrum of an element exposed to such an energy source consists of a collection of the allowable emission wavelengths, commonly called emission lines, because of the discrete nature of the emitted wavelengths. This emission spectrum can be used as a unique characteristic for qualitative identification of the element. Atomic emission using electrical arcs has been widely used in qualitative analysis. Emission techniques can also be used to determine how much of an element is present in a sample. For a "quantitative" analysis, the intensity of light emitted at the wavelength of the element to be determined is measured. The emission intensity at this wavelength will be greater as the number of atoms of the analyte element increases. The technique of flame photometry is an application of atomic emission for quantitative analysis.

Atomic Fluorescence Spectroscopy

The third field of atomic spectroscopy is atomic fluorescence. This technique incorporates aspects of both atomic absorption and atomic emission. Like atomic absorption, ground state atoms created in a flame are excited by focusing a beam of light into the atomic vapour. Instead of looking at the amount of light absorbed in the process, however, the emission resulting from the decay of the atoms excited by the source light is measured.

The intensity of this "fluorescence" increases with increasing atom concentration, providing the basis for quantitative determination. The source lamp for atomic fluorescence is mounted at an angle to the rest of the optical system, so that the light detector sees only the fluorescence in the flame and not the light from the lamp itself.

It is advantageous to maximise lamp intensity since sensitivity is directly related to the number of excited atoms which in turn is a function of the intensity of the exciting radiation. While atomic absorption is the most widely applied of the three techniques and usually offers several advantages over the other two, particular benefits may be gained with either emission or fluorescence in special analytical situations.

Andor Solutions for Atomic Spectroscopy

Considering the needs of atomic spectroscopy, Andor Technology has designed its line of products to fully cater to the needs of this industry.

Traditionally this application was used to using a combination of a monochromator and a PMT, however the solution that Andor provides consists of a high throughput (F/4 aperture ratio) imaging spectrograph ShamrockSR-303i coupled to an iStar ICCD camera. The iStar comes with a variety of photocathode tubes which can be selected depending on the range of wavelength desired to be detected. These cameras have a single photon sensitivity as found in PMTs and can be gated down to 2ns gate widths. The advantage of using this combination of an imaging spectrograph and ICCD detector, over the monochromator and PMT, is that a number of wavelengths can be detected in one shot, this drastically reduces the experiment time while simultaneously provides for high spatial and spectral resolution. Single photon sensitivity is offered by the gain multiplication ability of the intensifier tube.

Molecular Spectroscopy

Molecular spectroscopy deals with the response (spectrum) of molecules interacting with probing signals of known energy (or frequency, according to Planck's formula). Molecules have quantized energy levels that can be analysed by detecting the molecule's energy exchange through absorbance or emission. Spectroscopy does not generally refer to diffraction studies where particles such as neutrons, electrons, or high energy X-rays interact with a regular arrangement of molecules (as in a crystal).

Theoretical Aspects

The study of molecules by molecular physics and theoretical chemistry is largely based on quantum mechanics and is essential for the understanding of the chemical bond. The simplest of molecules is the hydrogen molecule-ion, H_2^+, and the simplest of all the chemical bonds is the one-electron bond. H_2^+ is composed of two positively charged protons and one negatively charged electron, which means that the Schrödinger equation for the system can be solved more easily due to the lack of electron–electron repulsion. With the development of fast digital computers, approximate solutions for more complicated molecules became possible and are one of the main aspects of computational chemistry.

When trying to define rigorously whether an arrangement of atoms is "sufficiently stable" to be considered a molecule, IUPAC suggests that it "must correspond to a depression on the potential energy surface that is deep enough to confine at least one vibrational state". This definition does not depend on the nature of the interaction between the atoms, but only on the strength of the interaction. In fact, it includes weakly bound species that would not traditionally be considered molecules, such as the helium dimer, He_2, which has one vibrational bound state and is so loosely bound that it is only likely to be observed at very low temperatures.

Chapter 3

Rotational Spectroscopy

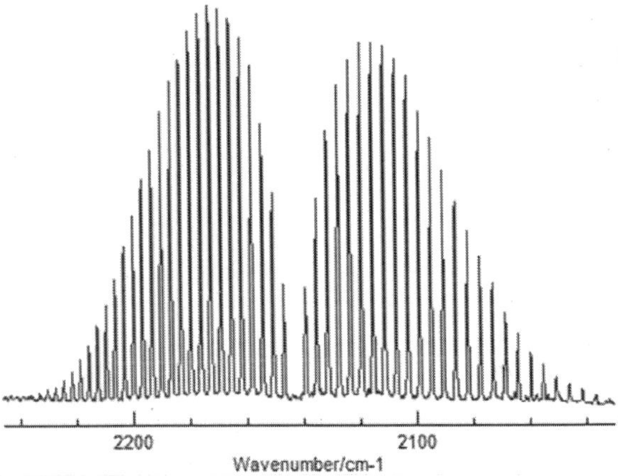

Figure: Part of the rotational-vibrational spectrum of carbon monoxide (CO) gas (from FTIR), showing the presence of P- and R- branches. Frequency is on the x-axis, and absorbance on the y-axis.

Rotational spectroscopy or microwave spectroscopy studies the absorption and emission of electromagnetic radiation (typically in the microwave region of the electromagnetic spectrum) by molecules associated with a corresponding change in the rotational quantum number of the molecule. The use of microwaves in spectroscopy essentially became possible due to the development of microwave technology for RADAR during World War II. Rotational spectroscopy is practical only in the gas phase where the rotational motion is quantized. In solids or liquids the rotational motion is usually quenched due to collisions.

Rotational spectrum from a molecule (to first order) requires that the molecule have a permanent dipole moment and that there be a difference between its centre of charge and its centre of mass, or equivalently a separation between two unlike charges. It is this dipole moment that enables the electric field of the radiation (microwave) to exert a torque on the molecule, causing it to rotate more quickly (in excitation) or slowly (in de-excitation). Homonuclear diatomic molecules such as dioxygen (O_2), dihydrogen (H_2), etc. do not have a dipole moment and, hence, no purely rotational spectrum. However, electronic excitations can lead to asymmetric charge distributions and thus provide a net dipole moment to the molecule. Under such circumstances, these molecules will exhibit a rotational spectrum.

Among the diatomic molecules, carbon monoxide (CO) has one of the simplest rotational spectra. As for tri-atomic molecules, hydrogen cyanide (HCa–N) has a simple rotational spectrum for a linear molecule and hydrogen isocyanide (HN=C:) for a non-linear molecule. As the number of atoms increases, the spectrum becomes more complex, as lines, due to different transitions, start overlapping.

Understanding the Rotational Spectrum

In quantum mechanics the free rotation of a molecule is quantized, that is the rotational energy and the angular momentum can take only certain fixed values; what these values are is related simply to the moment of inertia, *I*, of the molecule. In general, for any molecule, there are three moments of inertia: I_A, *IB and* I_C about three mutually orthogonal axes *A*, *B*, and *C* with the origin at the centre of mass of the system. Linear molecules are a special case because one of the moments of inertia (, the moment of inertia along molecular axis) is negligible (i.e., $I_A \ll I_B = I_C$).

The general convention is to define the axes such that the axis *A* has the smallest moment of inertia, the axis *B* axis is the next smallest and so on such that $I_A \leq I_B \leq I_C$. (A rarely used convention defines the axis along a symmetric axis of the molecule, if one exists. To avoid confusion, this chapter uses the prior convention.) The particular pattern of energy levels (and, hence, of transitions in the rotational spectrum) for a molecule is determined by its symmetry. A convenient way to look at the molecules is to divide them into four different classes (based on the symmetry of their structure). These are linear molecules, symmetric tops, spherical tops, and asymmetric tops. (Equivalently, many authors use the word "rotor" instead of "top".)

Linear Molecules

For a linear molecule the moments of inertia are related by

$I_A \ll I_B = I_C$. For most of the purposes, I_A is taken to be zero. For a linear molecule, the separation of lines in the rotational spectrum can be related directly to the moment of inertia of the molecule, and, for a molecule of known atomic masses, can be used to determine the bond lengths (structure) directly. For diatomic molecules, this process is trivial, and can be made from a single measurement of the rotational spectrum. For linear molecules with more atoms, rather more work is required, and it is necessary to measure molecules in which more than one isotope of each atom have been substituted (effectively this gives rise to a set of simultaneous equations that can be solved for the bond lengths). Examples of linear molecules: dioxygen (O=O), carbon monoxide (Oa–C*), hydroxy radical (OH), carbon dioxide (O=C=O), hydrogen cyanide (HCa–N), carbonyl sulfide (O=C=S), chloroethyne (HCa–CCl), acetylene (HCa–CH)

Symmetric Tops

A symmetric top is a molecule in which two moments of inertia are the same. As a matter of historical convenience, spectroscopists divide molecules into two classes of symmetric tops, *Oblate symmetric tops* (saucer or disc shaped) with $I_A = I_B < I_C$ and *Prolate symmetric tops* (rugby football, or cigar shaped) with $I_A < I_B = I_C$. The spectra look rather different, and are instantly recognisable. As with linear molecules, the structure of *symmetric tops* (bond lengths and bond angles) can be deduced from their spectra.

Examples of symmetric tops:

Oblate: benzene (C_6H_6), cyclobutadiene (C_4H_4), ammonia (NH_3)

Prolate: chloromethane (CH_3Cl), propyne (CH_3Ca–CH)

As a detailed example, ammonia has a moment of inertia I_C = 4.4128×10^{-47} kg m^2 about the threefold symmetry axis (C_3), and moments $I_A = I_B = 2.8059 \times 10^{-47}$ kg m^2 about any axis perpendicular to the C_3 axis. Since the unique moment of inertia is larger than the other two, the molecule is an oblate symmetric top.

Spherical Tops

A spherical top molecule can be considered as a special case of *symmetric tops* with equal moment of inertia about all three axes ($I_A = I_B = I_C$). Examples of spherical tops: phosphorus tetramer (P_4), carbon tetrachloride (CCl_4), methane (CH_4), ammonium ion (NH_4^+), sulfur hexafluoride (SF_6).

Asymmetric Tops

An asymmetric top occurs if all three moments of inertia are different. Most large molecules are asymmetric tops, even when they have a high degree of symmetry. In general, for such molecules, a simple interpretation of the spectrum is not normally possible. However, molecules have spectra similar to those of a linear molecule or a symmetric top, in which case the molecular structure may also be similar to that of a linear molecule or a symmetric top. For the most general case, however, all that can be done is to fit the spectra to three different moments of inertia. If the molecular formula is known, then educated guesses can be made of the possible structure, and, from this guessed structure, the moments of inertia can be calculated. If the calculated moments of inertia agree well with the measured moments of inertia, then the structure can be said to have been determined. For this approach to determining molecular structure, isotopic substitution is invaluable. Examples of asymmetric tops: anthracene ($C_{14}H_{10}$), water (H_2O), nitrogen dioxide (NO_2)

Structure of Rotational Spectra

Linear Molecules

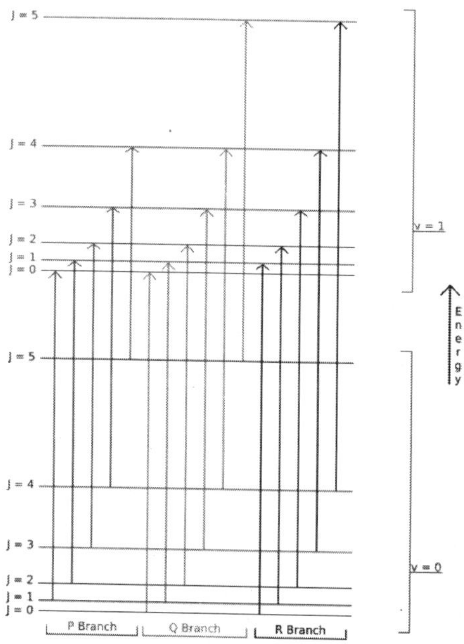

Figure:

An energy level diagram showing some of the transitions involved in the IR vibration-rotation spectrum of a linear molecule: P branch (where $\Delta J = -1$), Q branch (not always allowed, $\Delta J = 0$) and R branch ($\Delta J = 1$)

These molecules have two degenerate modes of rotation $(I_B = I_C, I_A = 0)$. Since we cannot distinguish between the two modes, we need only one rotational quantum number (J) to describe the rotational motion of the molecule.

The rotational energy levels ($F(J)$) of the molecule based on rigid rotor model can be expressed as,

$$F(J)F(J) = \tilde{B}_e J(J+1) \qquad J = 0,1,2,...$$

where $\tilde{B}_e$ is the rotational constant of the molecule and is related to the moment of inertia of the molecule $I_B = I_C$ by

$$\tilde{B}_e = \frac{h}{8\pi^2 c I_B}$$

(in wavenumber scale).

Selection rules dictate that during emission or absorption the rotational quantum number has to change by unity; i.e., $\Delta J = J' - J'' = \pm 1$. Thus, the locations of the lines in a rotational spectrum will be given by

$$\tilde{\nu}_{J' \leftrightarrow J''} = F(J') - F(J'') = 2\tilde{B}_e(J''+1) \qquad J'' = 0,1,2,...$$

where J'' denotes the lower energy level and J' denotes higher energy level involved in the transition. The height of the lines is determined by the distribution of the molecules in the different levels and the probability of transition between two energy levels.

We observe that, for a rigid rotor, the transition lines are equally spaced in the wavenumber space. However, this is not always the case, except for the rigid rotor model. For non-rigid rotor model, we need to consider changes in the moment of inertia of the molecule. Two primary reasons for this are.

Centrifugal Distortion

When a molecule rotates, the centrifugal force pulls the atoms apart. As a result, the moment of inertia of the molecule increases, thus decreasing $\tilde{B}_e$. To account for this a centrifugal distortion correction term is added to the rotational energy levels of the molecule.

$$F(J) = \tilde{B}_e J(J+1) - \tilde{D}_e J^2 (J+1)^2 \qquad J = 0,1,2,...$$

where $\tilde{D}_e$ is the centrifugal distortion constant.

Therefore, the line spacing for the rotational mode changes to,

$$\tilde{\nu}_{J' \leftrightarrow J''} = 2\tilde{B}_e (J''+1) - 4\tilde{D}_e (J''+1)^3 \qquad J'' = 0,1,2,...$$

Effect of Vibration on Rotation

A molecule is always in vibration. As the molecule vibrates, its moment of inertia changes. Further, there is a fictitious force, Coriolis coupling, between the vibrational motion of the nuclei in the rotating (non-inertial) frame. However, as long as the vibrational quantum number does not change (i.e., the molecule is in only one state of vibration), the effect of vibration on rotation is not important, because the time for vibration is much shorter than the time required for rotation. The Coriolis coupling is often negligible, too, if one is interested in low vibrational and rotational quantum numbers only.

Symmetric Top

The rotational motion of a symmetric top molecule can be described by two independent rotational quantum numbers (since two axes have equal moments of inertia, the rotational motion about these axes requires only one rotational quantum number for complete description).

Instead of defining the two rotational quantum numbers for two independent axes, we associate one of the quantum number (*J*) with the total angular momentum of the molecule and the other quantum number (*K*) with the angular momentum of the axis that has different moment of inertia (i.e., axis *C* for oblate symmetric top and axis *A* for prolate symmetric tops). The rotational energy $F(J,K)$ of such a molecule, based on rigid rotor assumptions can be expressed in terms of the two previously defined rotational quantum numbers as follows

$$F(J,K) = \tilde{B}J(J+1) + (\tilde{A} - \tilde{B})K^2 \qquad J = 0,1,2,...$$

and $K = -J, -J+1, ..., -1, 0, 1, ..., J-1, J$

where $\tilde{B} = \dfrac{h}{8\pi^2 c I_B}$ and $\tilde{A} = \dfrac{h}{8\pi^2 c I_A}$ for a *prolate* symmetric top molecule or $\tilde{A} = \dfrac{h}{8\pi^2 c I_C}$ for an *oblate* molecule.

Selection rule for these molecules provide the guidelines for possible transitions. Therefore,

$$\Delta J = \pm 1 \quad \text{and} \quad \Delta K = 0.$$

This is so because K is associated with the axis about which the molecule is symmetric and, hence, has no net dipole moment in that direction. Thus, there is no interaction of this mode with the light particles (photons).

This gives the transition wavenumbers of

$$\tilde{\nu}_{J' \leftrightarrow J'',K} = F(J',K) - F(J'',K) = 2\tilde{B}(J''+1) \qquad J'' = 0,1,2,\ldots$$

which is the same as in the case of a linear molecule.

In case of non-rigid rotors, the first order centrifugal distortion correction is given by

$$F(J,K) = \tilde{B}J(J+1) + (\tilde{A}-\tilde{B})K^2 - \tilde{D}_J J^2 (J+1)^2 - \tilde{D}_{JK} J(J+1)K^2 - D_K K^4$$

$$J = 0,1,2,\ldots \quad \text{and} \quad K = -J,\ldots,0,\ldots,J$$

The suffixes on the centrifugal distortion constant D indicate the rotational mode involved and are not a function of the rotational quantum number. The location of the transition lines on a spectrum is given by

$$\tilde{\nu}_{J' \leftrightarrow J'',K} = F(J',K) - F(J'',K) = 2\tilde{B}(J''+1) - 4D_J (J''+1)^3 - 2D_{JK} (J''+1)K^2$$

$$J'' = 0,1,2,\ldots$$

Spherical Top

Unlike other molecules, spherical top molecules have no net dipole moment, and, hence, they do not exhibit a pure rotational spectrum.

Asymmetric Top

The spectrum for these molecules usually involves many lines due to three different rotational modes and their combinations. The following analysis is valid for the general case and collapses to the various special cases described above in the appropriate limit.

From the moments of inertia one can define an asymmetry parameter k as

$\kappa = \dfrac{2B - A - C}{A - C}$, which varies from -1 for a prolate symmetric top to 1 for an oblate symmetric top.

One can define a scaled rotational Hamiltonian dependent on J and κ. The (symmetric) matrix representation of this Hamiltonian is banded, zero everywhere but the main diagonal and the second subdiagonal. The Hamiltonian can be formulated in six different settings, dependent on the mapping of the principal axes to lab axes and handedness. For the most asymmetric, right-handed representation, the diagonal elements are, for $|k| \leq J$

$$H_{k,k}(\kappa) = \kappa k^2$$

and the second off-diagonal elements (independent of κ) are

$$H_{k,k+2}(\kappa) = \sqrt{[J(J+1)-(k+1)(k+2)][J(J+1)-k(k+1)]}\,/\,2$$

Diagonalising H yields a set of $2J+1$ scaled rotational energy levels $E_k(\kappa)$. The rotational energy levels of the asymmetric rotor for total angular momentum J are then given by

$$\frac{A+C}{2}J(J+1)+\frac{A-C}{2}E_k(\kappa)$$

Hyperfine Interaction

In addition to the main structure that is observed in microwave spectra due to the rotational motion of the molecules, a whole host of further interactions are responsible for small details in the spectra, and the study of these details provides a very deep understanding of molecular quantum mechanics.

The main interactions responsible for small changes in the spectra (additional splittings and shifts of lines) are due to magnetic and electrostatic interactions in the molecule.

The particular strength of such interactions differs in different molecules, but, in general, the order of these effects (in decreasing significance) is:

1. *electron spin - electron spin interaction* (this occurs in molecules with two or more unpaired electrons, and is a magnetic-dipole/ magnetic-dipole interaction)
2. *electron spin - molecular rotation* (the rotation of a molecule corresponds to a magnetic dipole, which interacts with the magnetic dipole moment of the electron)
3. *electron spin - nuclear spin interaction* (the interaction between the magnetic dipole moment of the electron and the magnetic dipole moment of the nuclei (if present)).

4. *electric field gradient - nuclear electric quadrupole interaction* (the interaction between the electric field gradient of the electron cloud of the molecule and the electric quadrupole moments of nuclei (if present)).
5. *nuclear spin - nuclear spin interaction* (nuclear magnetic moments interacting with one another).

These interactions give rise to the characteristic energy levels that are probed in "magnetic resonance" spectroscopy such as NMR and ESR, where they represent the "zero field splittings," which are always present.

Experimental Determination of the Spectrum

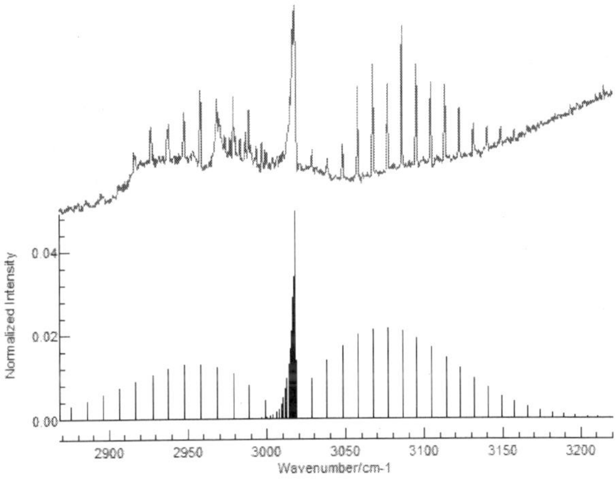

Figure: Part of the rotational-vibrational spectrum of methane (CH4) gas (from FTIR), showing the presence of P-, Q- and R- branches (purple, top) and a simulation in PGOPHER (black, bottom). Frequency is on the x-axis, and transmittance on the y-axis.

Fourier transform infrared (FTIR) spectroscopy can be used to experimentally study rotational spectra. Typical spectra at these wavelengths involve rovibrational excitation, i.e., excitation of both a vibrational and a rotational mode of a molecule.

In the past, microwave spectra were determined using a simple arrangement in which low-pressure gas was introduced to a section of waveguide between a microwave source (of variable frequency) and a microwave detector. The spectrum was obtained by sweeping the frequency of the source while detecting the intensity of the transmitted radiation. This experimental arrangement has a major difficulty related to the propagation of microwave radiation through waveguides. The physical size of the waveguide restricts the frequency of the radiation that can be transmitted through it. For a given waveguide size (such as X-band), there is a cutoff frequency, and microwave radiation with smaller frequencies (longer wavelengths) cannot be propagated through the waveguide. In addition, as the frequency is increased, additional modes of propagation become possible, which correspond to different velocities of the radiation propagating down the waveguide (this can be envisaged as the radiation bouncing down the guide, at different angles of reflection). The net result of these considerations is that each size of waveguide is useful only over a rather narrow range of frequencies and must be physically swapped out for a different size of waveguide once this frequency range is exceeded.

From 1980 onward, microwave spectra have often been obtained using Fourier Transform Microwave Spectroscopy - a technique developed by W. H. Flygare at the University of Illinois.

Within the last two years, a further development of Fourier Transform Microwave Spectroscopy has occurred, which may well introduce a new renaissance into microwave spectroscopy. This is the use of "Chirped Pulses" to provide an electromagnetic wave that has as its Fourier Transform a very wide range of microwave frequencies.

Rovibrational coupling is a coupled rotational and vibrational excitation of a molecule. It is different from rovibronic coupling, which involves a change in all of electronic, vibrational, and rotational states simultaneously.

Rotational-Vibrational Spectroscopy

Generally vibrational transitions occur in conjunction with rotational transitions. Consequently, it is possible to observe both rotational and vibrational transitions in the vibrational spectrum. Although many methods are available for observing vibrational spectra, the two most common methods are infrared spectroscopy and Raman spectroscopy. The energy of rotational transitions is on the order of 10^{-23} J whereas vibrational transitions have energies on the order of 10^{-20} J. Therefore, highly-resolved vibrational spectra will contain fine structure corresponding to the rotational transitions that occur at the same time as a vibrational transition. Although molecular vibrations and rotations do have some effect on one other, this interaction is usually small. Consequently, the rotational and vibrational contributions to the energy of the molecule can be considered independently to a first approximation:

$$E_{vib,rot} = E_{vib} + E_{rot} = \left(n + \frac{1}{2}\right)h\nu_0 + hc\bar{B}J\left(J+1\right)$$

where n is the vibrational quantum number, J is the rotational quantum number, h is Planck's constant, V_O is the frequency of the vibration, c is the speed of light, and $\bar{B}$ is the rotational constant.

Evaluating Spectra

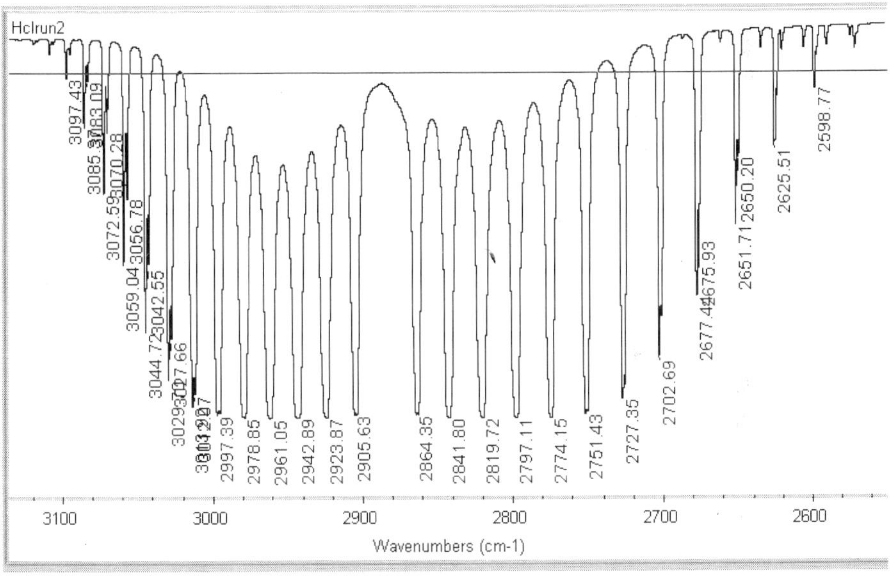

Figure: An energy level diagram showing some of the transitions involved in the IR vibration-rotation spectrum

The strict selection rule for the absorption of dipole radiation (the strongest component of light) is that $\Delta J = 0, \pm 1$. This is because of the vector addition properties of quantum mechanical angular momenta, and because light particles (photons) have angular momenta of 1. In spectroscopy, the transitions where $\Delta J = -1$ are referred to as the P-branch.

Transitions with $\Delta J = 0$ are referred to as Q-branch, and $\Delta J = 1$ as R-branch. For linear molecules the most commonly observed case is that only transitions with $\Delta J = \pm 1$ are observed. This is only possible when the molecule has a "singlet" ground state, that is there are no unpaired electron spins in the molecule. For molecules that do have unpaired electrons, Q branches are commonly observed.

The gap between the R- and P-branches is known as the Q-branch. A peak would appear here for a vibrational transition in which the rotational energy did not change ($\Delta J = 0$). However, according to the quantum mechanical rigid rotor model upon which rotational spectroscopy is based, there is a spectroscopic selection rule that requires that $\Delta J = \pm 1$. This selection rule explains why the P- and R-branches are observed, but not the Q-branch (as well as branches for which $\Delta J = \pm 2$, $\Delta J = \pm 3$, etc.).

The positions of the peaks in the spectrum can be predicted using the rigid rotor model. One prediction of the rigid rotor model is that the space between each of the peaks should be $2\bar{B}$ where $\bar{B}$ is the rotational constant for a given molecule. Experimentally, it is observed that the spacing between the R-branch peaks decreases as the frequency increases. Similarly, the spacing between the P-branch peaks increases as the frequency decreases. This variation in the spacing results from the bonds between the atoms in a molecule not being entirely rigid.

This variation can be mostly accounted for using a slightly more complex model that takes into account the variation in the rotational constant as the vibrational energy changes. Using this model, the positions of the R-branch peaks are predicted to be at:

$$\bar{\nu}_R = \nu_0 + 2\bar{B}_1 + \left(3\bar{B}_1 - \bar{B}_0\right)J + \left(\bar{B}_1 - \bar{B}_0\right)J^2 \qquad J = 0,1,2...$$

where $\bar{B}_0$ is the rotational constant for the $n = 0$ vibrational level and $\bar{B}_1$ is the rotational constant for the $n = 1$ vibrational level. Likewise, the P-branch peaks are predicted to be at:

$$\bar{\nu}_P = \bar{\nu}_0 - \left(\bar{B}_1 + \bar{B}_0\right)J + \left(\bar{B}_1 - \bar{B}_0\right)J^2 \qquad J = 1,2,3...$$

Rotational-vibrational spectra will also show some fine structure due to the presence of different isotopes in the spectrum. In the spectrum shown above, all of the rotational peaks are slightly split into two peaks. One peak corresponds to ^{35}Cl and the other to ^{37}Cl. The ratio of the peak intensities corresponds to the natural abundance of these two isotopes.

Applications: Microwave spectroscopy is commonly used in physical chemistry to determine the structure of small molecules (such as ozone, methanol, or water) with high precision. Other common techniques for determining molecular structure, such as X-ray crystallography do not work very well for some of these molecules (especially the gases) and are not as precise. However, microwave spectroscopy is not useful for determining the structures of large molecules such as proteins.

Modern microwave spectrometres have very high resolution. When hyperfine structure can be observed, the technique can also provide information on the electronic structures of molecules.

Microwave spectroscopy is one of the principal means by which the constituents of the universe are determined from the earth. It is particularly useful for detecting molecules in the interstellar medium (ISM). One of the early surprises in interstellar chemistry came with the discovery of the existence in the ISM of long-chain carbon molecules. It was in attempting to research such molecules in the laboratory that Harry Kroto was led to the laboratory of Rick Smalley and Robert Curl, where it was possible to vapourise carbon under enormous energy conditions. This collaborative experiment led to the discovery of C_{60}, buckminsterfullerene, which led to the award of the 1996 Nobel prize in chemistry to Kroto, Smalley and Curl.

Mechanical Properties of Rotation

Torque: m r

Moment of inertia, *I*: $I = \sum_i m_i r_i^2$

The moment of inertia of a molecule is a measure of how difficult it is to rotationally accelerate the molecule - the larger the moment of inertia, the smaller the increase in angular momentum for a given applied torque.

Because *I* depends on both the mass of the atoms and the geometry of the molecule, the rotational spectroscopy will provide us with information about *bond lengths and bond angles.*

Angular momentum: $J = I \cdot \omega$

where $\omega = 2\pi\nu$, the angular velocity of a rotating object with ν rotational frequency.

Chapter 4

Vibrational Spectra

Molecular Vibration

A molecular vibration occurs when atoms in a molecule are in periodic motion while the molecule as a whole has constant translational and rotational motion. The frequency of the periodic motion is known as a vibration frequency, and the typical frequencies of molecular vibrations range from less than 10^{12} to approximately 10^{14} Hz.

In general, a molecule with N atoms has $3N - 6$ normal modes of vibration, but a *linear* molecule has $3N - 5$ such modes, as rotation about its molecular axis cannot be observed. A diatomic molecule has one normal mode of vibration. The normal modes of vibration of polyatomic molecules are independent of each other but each normal mode will involve simultaneous vibrations of different parts of the molecule such as different chemical bonds.

A molecular vibration is excited when the molecule absorbs a quantum of energy, E, corresponding to the vibration's frequency, ν, according to the relation $E = h\nu$ (where h is Planck's constant). A fundamental vibration is excited when one such quantum of energy is absorbed by the molecule in its ground state. When two quanta are absorbed the first overtone is excited, and so on to higher overtones.

To a first approximation, the motion in a normal vibration can be described as a kind of simple harmonic motion. In this approximation, the vibrational energy is a quadratic function (parabola) with respect to the atomic displacements and the first overtone has twice the frequency of the fundamental.

In reality, vibrations are anharmonic and the first overtone has a frequency that is slightly lower than twice that of the fundamental. Excitation of the higher overtones involves progressively less and less additional energy and eventually leads to dissociation of the molecule, as the potential energy of the molecule is more like a Morse potential. The vibrational states of a molecule can be probed in a variety of ways.

The most direct way is through infrared spectroscopy, as vibrational transitions typically require an amount of energy that corresponds to the infrared region of the spectrum. Raman spectroscopy, which typically uses visible light, can also be used to measure vibration frequencies directly.

Vibrational excitation can occur in conjunction with electronic excitation (vibronic transition), giving vibrational fine structure to electronic transitions, particularly with molecules in the gas state. Simultaneous excitation of a vibration and rotations gives rise to vibration-rotation spectra.

Vibrational Coordinates

The coordinate of a normal vibration is a combination of *changes* in the positions of atoms in the molecule. When the vibration is excited the coordinate changes sinusoidally with a frequency ν, the frequency of the vibration.

Internal Coordinates

Internal coordinates are of the following types, illustrated with reference to the planar molecule ethylene,

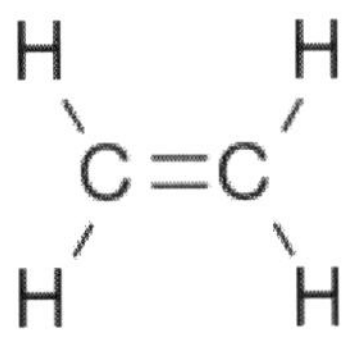

- Stretching: a change in the length of a bond, such as C-H or C-C
- Bending: a change in the angle between two bonds, such as the HCH angle in a methylene group
- Rocking: a change in angle between a group of atoms, such as a methylene group and the rest of the molecule.
- Wagging: a change in angle between the plane of a group of atoms, such as a methylene group and a plane through the rest of the molecule,

- Twisting: a change in the angle between the planes of two groups of atoms, such as a change in the angle between the two methylene groups.
- Out-of-plane: a change in the angle between any one of the C-H bonds and the plane defined by the remaining atoms of the ethylene molecule. Another example is in BF_3 when the boron atom moves in and out of the plane of the three fluorine atoms.

In a rocking, wagging or twisting coordinate the bond lengths within the groups involved do not change. The angles do. Rocking is distinguished from wagging by the fact that the atoms in the group stay in the same plane. In ethene there are 12 internal coordinates: 4 C-H stretching, 1 C-C stretching, 2 H-C-H bending, 2 CH_2 rocking, 2 CH_2 wagging, 1 twisting.

Note that the H-C-C angles cannot be used as internal coordinates as the angles at each carbon atom cannot all increase at the same time.

Vibrations of a Methylene Group (-CH_2-) in a Molecule for Illustration

The atoms in a CH_2 group, commonly found in organic compounds, can vibrate in six different ways: symmetric and antisymmetric stretching, scissoring, rocking, wagging and twisting as shown here:

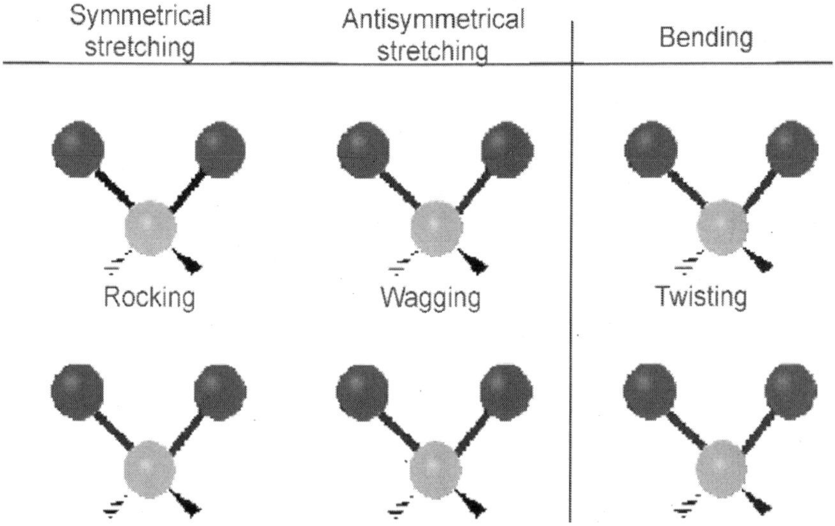

Figure: (These figures do not represent the "recoil" of the C atoms, which, though necessarily present to balance the overall movements of the molecule, are much smaller than the movements of the lighter H atoms).

Symmetry-Adapted Coordinates

Symmetry-adapted coordinates may be created by applying a projection operator to a set of internal coordinates. The projection operator is constructed with the aid of the character table of the molecular point group. For example, the four(un-normalised) C-H stretching coordinates of the molecule ethene are given by

$$Q_{s1} = q_1 + q_2 + q_3 + q_4$$

$$Q_{s2} = q_1 + q_2 - q_3 - q_4$$

$$Q_{s3} = q_1 - q_2 + q_3 - q_4$$

$$Q_{s4} = q_1 - q_2 - q_3 + q_4$$

where q_1 - q_4 are the internal coordinates for stretching of each of the four C-H bonds.

Illustrations of symmetry-adapted coordinates for most small molecules can be found in Nakamoto.

Normal Coordinates

The normal coordinates, denoted as Q, refer to the positions of atoms away from their equilibrium positions, with respect to a normal mode of vibration. Each normal mode is assigned a single normal coordinate, and so the normal coordinate refers to the "progress" along that normal mode at any given time. Formally, normal modes are determined by solving a secular determinant, and then the normal coordinates (over the normal modes) can be expressed as a summation over the cartesian cordinates (over the atom positions). The advantage of working in normal modes is that they diagonalize the matrix governing the molecular vibrations, so each normal mode is an independent molecular vibration, associated with its own spectrum of quantum mechanical states. If the molecule possesses symmetries, it will belong to a point group, and the normal modes will "transform as" an irreducible representation under that group. The normal modes can then be qualitatively determined by applying group theory and projecting the irreducible representation onto the cartesian coordinates. For example, when this treatment is applied to CO_2, it is found that the C=O stretches are not independent, but rather there is a O=C=O symmetric stretch and an O=C=O asymmetric stretch.

- symmetric stretching: the sum of the two C-O stretching coordinates; the two C-O bond lengths change by the same amount and the carbon atom is stationary. $Q = q_1 + q_2$

- asymmetric stretching: the difference of the two C-O stretching coordinates; one C-O bond length increases while the other decreases. $Q = q_1 - q_2$

When two or more normal coordinates belong to the same irreducible representation of the molecular point group (colloquially, have the same symmetry) there is "mixing" and the coefficients of the combination cannot be determined *a priori*. For example, in the linear molecule hydrogen cyanide, HCN, The two stretching vibrations are

1. principally C-H stretching with a little C-N stretching; $Q_1 = q_1 + a\ q_2$ (a << 1)
2. principally C-N stretching with a little C-H stretching; $Q_2 = b\ q_1 + q_2$ (b << 1)

The coefficients a and b are found by performing a full normal coordinate analysis by means of the Wilson GF method.

Newtonian Mechanics

Perhaps surprisingly, molecular vibrations can be treated using Newtonian mechanics to calculate the correct vibration frequencies. The basic assumption is that each vibration can be treated as though it corresponds to a spring. In the harmonic approximation the spring obeys Hooke's law: the force required to extend the spring is proportional to the extension. The proportionality constant is known as a *force constant, k*. The anharmonic oscillator is considered elsewhere.

$$\text{Force} = -kQ$$

By Newton's second law of motion this force is also equal to a reduced mass, μ, times acceleration.

$$\text{Force} = \mu \frac{d^2Q}{dt^2}$$

Since this is one and the same force the ordinary differential equation follows.

$$\mu \frac{d^2Q}{dt^2} + kQ = 0$$

The solution to this equation of simple harmonic motion is

$$Q(t) = A\cos(2\pi\nu t);\ \ \nu = \frac{1}{2\pi}\sqrt{\frac{k}{\mu}}$$

A is the maximum amplitude of the vibration coordinate Q. It remains to define the reduced mass, μ. In general, the reduced mass of a diatomic molecule, AB, is expressed in terms of the atomic masses, m_A and m_B, as

$$\frac{1}{\mu}=\frac{1}{m_A}+\frac{1}{m_B}.$$

The use of the reduced mass ensures that the centre of mass of the molecule is not affected by the vibration. In the harmonic approximation the potential energy of the molecule is a quadratic function of the normal coordinate. It follows that the force-constant is equal to the second derivative of the potential energy.

$$k=\frac{\partial^2 V}{\partial Q^2}$$

When two or more normal vibrations have the same symmetry a full normal coordinate analysis must be performed. The vibration frequencies, ν_i are obtained from the eigenvalues, λ_i, of the matrix product *GF*. *G* is a matrix of numbers derived from the masses of the atoms and the geometry of the molecule. *F* is a matrix derived from force-constant values. Details concerning the determination of the eigenvalues can be found in.

Quantum Mechanics

In the harmonic approximation the potential energy is a quadratic function of the normal coordinates. Solving the Schrödinger wave equation, the energy states for each normal coordinate are given by

$$E_n=h\left(n+\frac{1}{2}\right)\nu=h\left(n+\frac{1}{2}\right)\frac{1}{2\pi}\sqrt{\frac{k}{m}}$$

where n is a quantum number that can take values of 0, 1, 2 ... The difference in energy when n changes by 1 are therefore equal to the energy derived using classical mechanics. Knowing the wave functions, certain selection rules can be formulated. For example, for a harmonic oscillator transitions are allowed only when the quantum number n changes by one,

$$\Delta n=\pm 1$$

but this does not apply to an anharmonic oscillator; the observation of overtones is only possible because vibrations are anharmonic. Another consequence of anharmonicity is that transitions such as between states n=2 and n=1 have slightly less energy than transitions between the ground state and first excited state. Such a transition gives rise to a hot band.

Intensities

In an infrared spectrum the intensity of an absorption band is proportional to the derivative of the molecular dipole moment with respect to the normal coordinate. The intensity of Raman bands depends on polarizability.

Near-Infrared Spectroscopy

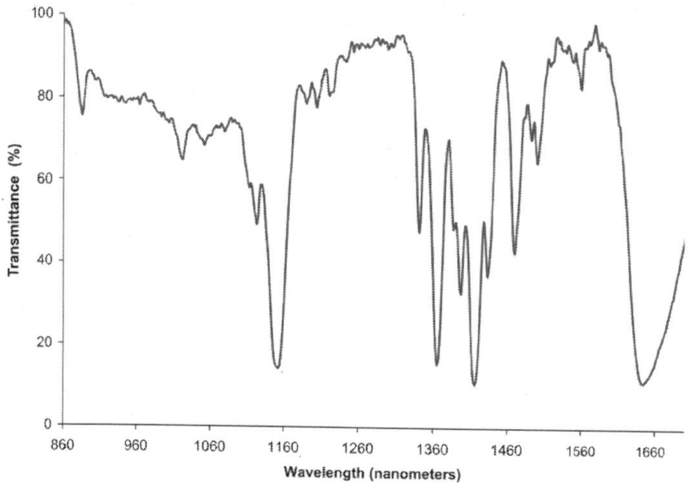

Figure: Near IR absorption spectrum of dichloromethane showing complicated overlapping overtones of mid IR absorption features.

Near-infrared spectroscopy (NIRS) is a spectroscopic method that uses the near-infrared region of the electromagnetic spectrum (from about 800 nm to 2500 nm). Typical applications include pharmaceutical, medical diagnostics (including blood sugar and pulse oximetry), food and agrochemical quality control, and combustion research, as well as research in functional neuroimaging, sports medicine & science, elite sports training, ergonomics, rehabilitation, neonatal research, brain computer interface, urology (bladder contraction) and neurology (neurovascular coupling).

Theory

Near-infrared spectroscopy is based on molecular overtone and combination vibrations. Such transitions are forbidden by the selection rules of quantum mechanics. As a result, the molar absorptivity in the near IR region is typically quite small. One advantage is that NIR can typically penetrate much farther into a sample than mid infrared radiation. Near-infrared spectroscopy is, therefore, not a particularly sensitive technique, but it can be very useful in probing bulk material with little or no sample preparation.

The molecular overtone and combination bands seen in the near IR are typically very broad, leading to complex spectra; it can be difficult to assign specific features to specific chemical components. Multivariate (multiple variables) calibration techniques (e.g., principal components analysis, partial least squares, or artificial neural networks) are often employed to extract the desired chemical information. Careful development of a set of calibration samples and application of multivariate calibration techniques is essential for near-infrared analytical methods.

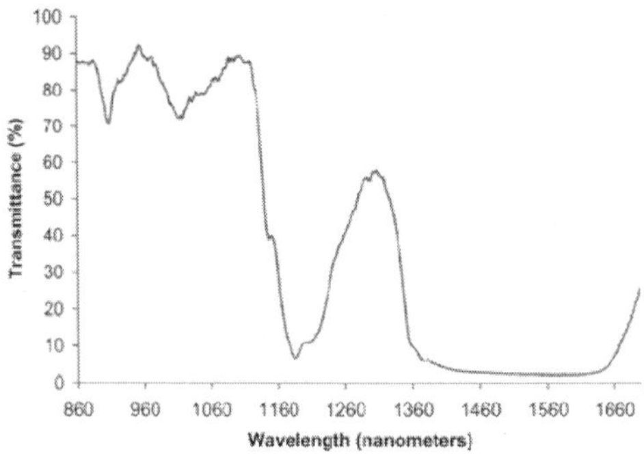

Figure: Near infrared spectrum of liquid ethanol.

The discovery of near-infrared energy is ascribed to Herschel in the 19th century, but the first industrial application began in the 1950s. In the first applications, NIRS was used only as an add-on unit to other optical devices that used other wavelengths such as ultraviolet (UV), visible (Vis), or mid-infrared (MIR) spectrometres. In the 1980s, a single unit, stand-alone NIRS system was made available, but the application of NIRS was focused more on chemical analysis. With the introduction of light-fibre optics in the mid-1980s and the monochromator-detector developments in early-1990s, NIRS became a more powerful tool for scientific research.

This optical method can be used in a number of fields of science including physics, physiology, or medicine. It is only in the last few decades that NIRS began to be used as a medical tool for monitoring patients.

Instrumentation

Instrumentation for near-IR (NIR) spectroscopy is similar to instruments for the UV-visible and mid-IR ranges. There is a source, a detector, and a dispersive element (such as a prism, or, more commonly, a diffraction grating) to allow the intensity at different wavelengths to be recorded. Fourier transform NIR instruments using an interferometre are also common, especially for wavelengths above ~1000 nm. Depending on the sample, the spectrum can be measured in either reflection or transmission.

Common incandescent or quartz halogen light bulbs are most often used as broadband sources of near-infrared radiation for analytical applications. Light-emitting diodes (LEDs) are also used; they offer greater lifetime and spectral stability and reduced power requirements.

The type of detector used depends primarily on the range of wavelengths to be measured. Silicon-based CCDs are suitable for the shorter end of the NIR range, but are not sufficiently sensitive over most of the range (over 1000 nm). InGaAs and PbS devices are more suitable though less sensitive than CCDs. In certain diode array (DA) NIRS instruments, both silicon-based and InGaAs detectors are employed in the same instrument. Such instruments can record both UV-visible and NIR spectra 'simultaneously'. Instruments intended for chemical imaging in the NIR may use a 2D array detector with a acousto-optic tunable filter. Multiple images may be recorded sequentially at different narrow wavelength bands.

Many commercial instruments for UV/vis spectroscopy are capable of recording spectra in the NIR range (to perhaps ~900 nm). In the same way, the range of some mid-IR instruments may extend into the NIR. In these instruments, the detector used for the NIR wavelengths is often the same detector used for the instrument's "main" range of interest.

Applications

The primary application of NIRS to the human body uses the fact that the transmission and absorption of NIR light in human body tissues contains information about hemoglobin concentration changes. When a specific area of the brain is activated, the localised blood volume in that area changes quickly. Optical imaging can measure the location and activity of specific regions of the brain by continuously monitoring blood hemoglobin levels through the determination of optical absorption coefficients. Typical applications of NIR spectroscopy include the analysis of foodstuffs, pharmaceuticals, combustion products and a major branch of astronomical spectroscopy.

Astronomical Spectroscopy

Near-infrared spectroscopy is used in astronomy for studying the atmospheres of cool stars where molecules can form. The vibrational and rotational signatures of molecules such as titanium oxide, cyanide, and carbon monoxide can be seen in this wavelength range and can give a clue towards the star's spectral type.

It is also used for studying molecules in other astronomical contexts, such as in molecular clouds where new stars are formed. The astronomical phenomenon known as reddening means that near-infrared wavelengths are less affected by dust in the interstellar medium, such that regions inaccessible by optical spectroscopy can be studied in the near-infrared. Since dust and gas are strongly associated, these dusty regions are exactly those where infrared spectroscopy is most useful. The near-infrared spectra of very young stars provide important information about their ages and masses, which is important for understanding star formation in general.

Remote Monitoring

Techniques have been developed for NIR spectroscopic imaging. Hyperspectral imaging has been applied for a wide range of uses, including the remote investigation of plants and soils. Data can be collected from instruments on airplanes or satellites to assess ground cover and soil chemistry.

Materials Science

Techniques have been developed for NIR spectroscopy of microscopic sample areas for film thickness measurements, research into the optical characteristics of nanoparticles and optical coatings for the telecommunications industry.

Medical Uses

Medical applications of NIRS centre on the non-invasive measurement of the amount and oxygen content of hemoglobin, as well as the use of exogenous optical tracers in conjunction with flow kinetics. NIRS can be used for non-invasive assessment of brain function through the intact skull in human subjects by detecting changes in blood hemoglobin concentrations associated with neural activity, e.g., in branches of Cognitive psychology as a partial replacement for fMRI techniques.

NIRS can be used on infants, and NIRS is much more portable than fMRI machines, even wireless instrumentation is available, which enables investigations in freely moving subjects. However, NIRS cannot fully replace fMRI because it can only be used to scan cortical tissue, where fMRI can be used to measure activation throughout the brain. Special public domain statistical toolboxes for analysis of stand alone and combined NIRS/MRI measurement have been developed (NIRS-SPM).

The application in functional mapping of the human cortex is called optical topography (OT), near infrared imaging (NIRI) or functional NIRS (fNIRS). The term optical tomography is used for three-dimensional NIRS. The terms NIRS, NIRI and OT are often used interchangeably, but they have some distinctions. The most important difference between NIRS and OT/NIRI is that OT/NIRI is used mainly to detect changes in optical properties of tissue simultaneously from multiple measurement points and display the results in the form of a map or image over a specific area, whereas NIRS provides quantitative data in absolute terms on up to a few specific points. The latter is also used to investigate other tissues such as, e.g., muscle, breast and tumours. NIRS can be used to quantify blood flow, blood volume, oygenation consumption, reoxygenation rates and muscle recovery time in muscle.

By employing several wavelengths and time resolved (frequency or time domain) and/or spatially resolved methods blood flow, volume and absolute tissue saturation ((StO_2 or Tissue Saturation Index (TSI)) can be quantified.. Applications of oximetry by NIRS methods include neuroscience, ergonomics, rehabilitation, brain computer interface, urology, the detection of illnesses which affect the blood circulation (e.g., peripheral vascular disease), the detection and assessment of breast tumors, and the optimization of training in sports medicine.

The use of NIRS in conjunction with a bolus injection of indocyanine green (ICG) has been used to measure cerebral blood flow and cerebral metabolic rate of oxygen consumption. It has also been shown that CMRO2 can be calculated with combined NIRS/MRI measurements. NIRS is starting to be used in pediatric critical care, to help deal with cardiac surgery post-op. Indeed, NIRS is able to measure venous oxygen saturation (SVO2), which is determined by the cardiac output, as well as other parameters (FiO2, hemoglobin, oxygen uptake). Therefore, following the NIRS gives critical care physicians a notion of the cardiac output. NIRS is liked by patients, because it is non-invasive, is painless, and uses non-ionizing radiation. The instrumental development of NIRS/NIRI/OT has proceeded tremendously during the last years and, in particular, in terms of quantification, imaging and miniaturization.

Particle Measurement

NIR is often used in particle sizing in a range of different fields, including studying pharmaceutical and agricultural powders.

Industrial Uses: As opposed to NIRS used in optical topography, general NIRS used in chemical assays does not provide imaging by mapping. For example, a clinical carbon dioxide analyser requires reference techniques and calibration routines to be able to get accurate CO_2 content change. In this case, calibration is performed by adjusting the zero control of the sample being tested after purposefully supplying 0% CO_2 or another known amount of CO_2 in the sample. Normal compressed gas from distributors contains about 95% O_2 and 5% CO_2, which can also be used to adjust %CO_2 metre reading to be exactly 5% at initial calibration.

GF Method

The GF method, sometimes referred to as FG method, is a classical mechanical method introduced by E. Bright Wilson to obtain certain *internal coordinates* for a vibrating semi-rigid molecule, the so-called *normal coordinates* Q_k. Normal coordinates decouple the classical vibrational motions of the molecule and thus give an easy route to obtaining vibrational amplitudes of the atoms as a function of time. In Wilson's GF method it is assumed that the molecular kinetic energy consists only of harmonic vibrations of the atoms, *i.e.,* overall rotational and translational energy is ignored. Normal coordinates appear also in a quantum mechanical description of the vibrational motions of the molecule and the Coriolis coupling between rotations and vibrations.

It follows from application of the Eckart conditions that the matrix G^{-1} gives the kinetic energy in terms of arbitrary linear internal coordinates, while F represents the (harmonic) potential energy in terms of these coordinates. The GF method gives the linear transformation from general internal coordinates to the special set of normal coordinates.

The GF Method

A non-linear molecule consisting of N atoms has $3N$-6 internal degrees of freedom, because positioning a molecule in three-dimensional space requires three degrees of freedom and the description of its orientation in space requires another three degree of freedom. These degrees of freedom must be subtracted from the $3N$ degrees of freedom of a system of N particles.

The atoms in a molecule are bound by a potential energy surface (PES) (or a force field) which is a function of $3N$-6 coordinates. The internal degrees of freedom q_1, ..., $q_{3N\text{-}6}$ describing the PES in an optimum way are often non-linear; they are for instance *valence coordinates*, such as bending and torsion angles and bond stretches. It is possible to write the quantum mechanical kinetic energy operator for such curvilinear coordinates, but it is hard to formulate a general theory applicable to any molecule.

This is why Wilson linearized the internal coordinates by assuming small displacements. The linearized version of the internal coordinate q_t is denoted by S_t.

The PES V can be Taylor expanded around its minimum in terms of the S_t. The third term (the Hessian of V) evaluated in the minimum is a force derivative matrix F. In the harmonic approximation the Taylor series is ended after this term. The second term, containing first derivatives, is zero because it is evaluated in the minimum of V. The first term can be included in the zero of energy. Thus,

$$2V \approx \sum_{s,t=1}^{3N-6} F_{st} S_s S_t \,.$$

The classical vibrational kinetic energy has the form:

$$2T = \sum_{s,t=1}^{3N-6} g_{st}(\mathbf{q}) \dot{S}_s \dot{S}_t ,$$

where g_{st} is an element of the metric tensor of the internal (curvilinear) coordinates. The dots indicate time derivatives. Mixed terms $S_s \dot{S}_t$ generally present in curvilinear coordinates are not present here, because only linear coordinate transformations are used.

Evaluation of the metric tensor g in the minimum q^0 of V gives the positive definite and symmetric matrix $G = g(q^0)^{-1}$. One can solve the following two matrix problems simultaneously

$$\mathbf{L}^T\mathbf{F}\mathbf{L} = \Phi \quad \text{and} \quad \mathbf{L}^T\mathbf{G}^{-1}\mathbf{L} = \mathbf{E},$$

since they are equivalent to the generalized eigenvalue problem

$$\mathbf{G}\mathbf{F}\mathbf{L} = \mathbf{L}\Phi,$$

where $\Phi = \text{diag}(f_1,\ldots,f_{3N-6})$ where f_i is equal to $4\pi^2\nu_i^2$ (ν_i is the frequency of normal mode i); E is the unit matrix. The matrix L^{-1} contains the *normal coordinates* Q_k in its rows:

$$Q_k = \sum_{t=1}^{3N-6} (\mathbf{L}^{-1})_{kt} S_t, \quad k = 1,\ldots,3N-6.$$

Because of the form of the generalized eigenvalue problem, the method is called the GF method, often with the name of its originator attached to it: Wilson's GF method. By matrix transposition in both sides of the equation and using the fact that both G and F are symmetric matrices, as are diagonal matrices, one can recast this equation into a very similar one for FG . This is why the method is also referred to as Wilson's FG method.

We introduce the vectors

$$\mathbf{s} = \text{col}(S_1,\ldots,S_{3N-6}) \quad \text{and} \quad \mathbf{Q} = \text{col}(Q_1,\ldots,Q_{3N-6}),$$

which satisfy the relation

$$\mathbf{s} = \mathbf{L}\mathbf{Q}.$$

Upon use of the results of the generalized eigenvalue equation, the energy $E = T + V$ (in the harmonic approximation) of the molecule becomes,

$$\begin{aligned} 2E &= \dot{\mathbf{s}}^T \mathbf{G}^{-1} \dot{\mathbf{s}} + \mathbf{s}^T\mathbf{F}\mathbf{s} \\ &= \dot{\mathbf{Q}}^T \left(\mathbf{L}^T\mathbf{G}^{-1}\mathbf{L}\right)\dot{\mathbf{Q}} + \mathbf{Q}^T\left(\mathbf{L}^T\mathbf{F}\mathbf{L}\right)\mathbf{Q} \\ &= \dot{\mathbf{Q}}^T \dot{\mathbf{Q}} + \mathbf{Q}^T\Phi\mathbf{Q} = \sum_{t=1}^{3N-6}\left(\dot{Q}_t^2 + f_t Q_t^2\right). \end{aligned}$$

The Lagrangian $L = T - V$ is

$$L = \frac{1}{2}\sum_{t=1}^{3N-6}\left(\dot{Q}_t^2 - f_t Q_t^2\right).$$

The corresponding Lagrange equations are identical to the Newton equations

$$\ddot{Q}_t + f_t Q_t = 0$$

for a set of uncoupled harmonic oscillators. These ordinary second-order differential equations are easily solved, yielding Q_t as a function of time.

Normal Coordinates in Terms of Cartesian Displacement Coordinates

Often the normal coordinates are expressed as linear combinations of Cartesian displacement coordinates. Let R_A be the position vector of nucleus A and $R_A{}^0$ the corresponding equilibrium position. Then $\mathbf{d}_A \equiv \mathbf{R}_A - \mathbf{R}_A^0$ is by definition the *Cartesian displacement coordinate* of nucleus A. Wilson's linearizing of the internal curvilinear coordinates q_t expresses the coordinate S_t in terms of the displacement coordinates

$$S_t = \sum_{A=1}^{N}\sum_{i=1}^{3} s_{Ai}^t d_{Ai} = \sum_{A=1}^{N} \mathbf{s}_A^t \cdot \mathbf{d}_A, \quad \text{for} \quad t = 1,\ldots,3N-6,$$

where $s_A{}^t$ is known as a *Wilson s-vector*. If we put the s_{Ai}^t into a $3N$-6 x $3N$ matrix B, this equation becomes in matrix language

$$\mathbf{s} = \mathbf{Bd}.$$

The actual form of the matrix elements of B can be fairly complicated. Especially for a torsion angle, which involves 4 atoms, it requires tedious vector algebra to derive the corresponding values of the . Now,

$$\mathbf{Q} = \mathbf{L}^{-1}\mathbf{s} = \mathbf{L}^{-1}\mathbf{Bd} \equiv \mathbf{Dd}.$$

In summation language:

$$Q_k = \sum_{A=1}^{N}\sum_{i=1}^{3} D_{Ai}^k d_{Ai} \quad \text{for} \quad k = 1,\ldots,3N-6.$$

Here D is a $3N$-6 x $3N$ matrix which is given by (i) the linearization of the internal coordinates q (an algebraic process) and (ii) solution of Wilson's GF equations (a numeric process).

Relation with Eckart Conditions

From the invariance of the internal coordinates S_t under overall rotation and translation of the molecule, follows the same for the linearized coordinates $s^t{}_A$. It can be shown that this implies that the following 6 conditions are satisfied by the internal coordinates,

$$\sum_{A=1}^{N} \mathbf{s}_A^t = 0 \quad \text{and} \quad \sum_{A=1}^{N} \mathbf{R}_A^0 \times \mathbf{s}_A^t = 0, \quad t = 1,\ldots,3N-6.$$

These conditions follow from the Eckart conditions that hold for the displacement vectors,

$$\sum_{A=1}^{N} M_A \, \mathbf{d}_A = 0 \quad \text{and} \quad \sum_{A=1}^{N} M_A \, \mathbf{R}_A^0 \times \mathbf{d}_A = 0.$$

Fermi Resonance

A Fermi resonance is the shifting of the energies and intensities of absorption bands in an infrared or Raman spectrum. It is a consequence of quantum mechanical mixing. The phenomenon was explained by the Italian physicist Enrico Fermi.

Selection Rules and Occurrence

Two conditions must be satisfied for the occurrence of Fermi Resonance:

- the two vibrational states of a molecule transform according to the same irreducible representation of the molecular point group. In other words the two vibrations must have the same symmetries (Mulliken symbols).
- The transitions (accidentally) have almost the same energy.

Fermi resonance most often occurs between normal and overtone modes, if they are nearly coincident in energy.

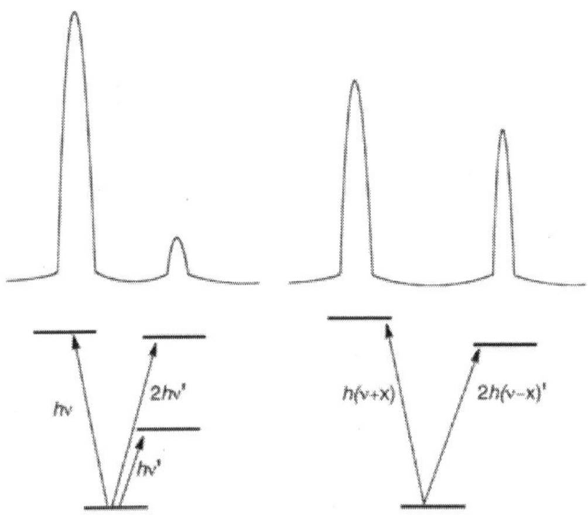

Figure: Ideallized appearance of a normal mode and an overtone before and after Fermi resonance. Beneath the idealized spectra are idealized energy level schemes.

Fermi resonance leads to two effects.

First, the high energy mode shifts to higher energy and the low energy mode shifts to still lower energy.

Second, the weaker mode gains intensity (becomes more allowed) and the more intense band decreases in intensity.

The two transitions are describable as a linear combination of the parent modes. Fermi resonance does not really lead to additional bands in the spectrum.

Examples

Ketones

High resolution IR spectra of most ketones reveal that the "carbonyl band– is split into a doublet. The peak separation is usually only a few cm^{-1}. This splitting arises from the mixing of ν_{CO} and the overtone of HCH bending modes.

CO_2

In CO_2, the bending vibration ν_2 (667 cm^{-1}) has symmetry Π_u. The first excited state of ν_2 is denoted 01^10 (no excitation in ν_1, one quantum of excitation in ν_2 with angular momentum about the molecular axis equal to ±1, no excitation in ν_3) and clearly transforms according to the irreducible representation Π_u.

Putting two quanta into ν_2 leads to a state with components of symmetry $(\Pi_u \times \Pi_u)_+ = \Sigma^+_g + \Delta_g$.

These are called 02^00 and 02^20, respectively. 02^00 has the same symmetry (Σ^+_g) as, and a very similar energy to, state 100 (a single excitation in the ν_1 symmetric stretch (calculated unperturbed frequency 1337 cm^{-1}), no excitation in ν_2, no excitation in ν_3). The states 02^00 and 100 can therefore mix together, producing a splitting and also a significant increase in the intensity of the 02^00 transition, so that both the 02^00 and 100 transitions have similar intensities.

Lennard-Jones Potential

The Lennard-Jones potential (also referred to as the L-J potential, 6-12 potential, or 12-6 potential) is a mathematically simple model that approximates the interaction between a pair of neutral atoms or molecules. A form of the potential was first proposed in 1924 by John Lennard-Jones. The most common expressions of the L-J potential are

$$V_{LJ} = 4\varepsilon \left[\left(\frac{\sigma}{r}\right)^{12} - \left(\frac{\sigma}{r}\right)^{6}\right]$$

$$= \varepsilon \left[\left(\frac{r_m}{r}\right)^{12} - 2\left(\frac{r_m}{r}\right)^{6}\right]$$

where ε is the depth of the potential well, σ is the finite distance at which the inter-particle potential is zero, r is the distance between the particles, and r_m is the distance at which the potential reaches its minimum. At r_m, the potential function has the value •ε. The distances are related as $r_m = 2^{1/6}\sigma$. These parameters can be fitted to reproduce experimental data or accurate quantum chemistry calculations. Due to its computational simplicity, the Lennard-Jones potential is used extensively in computer simulations even though more accurate potentials exist.

Explanation

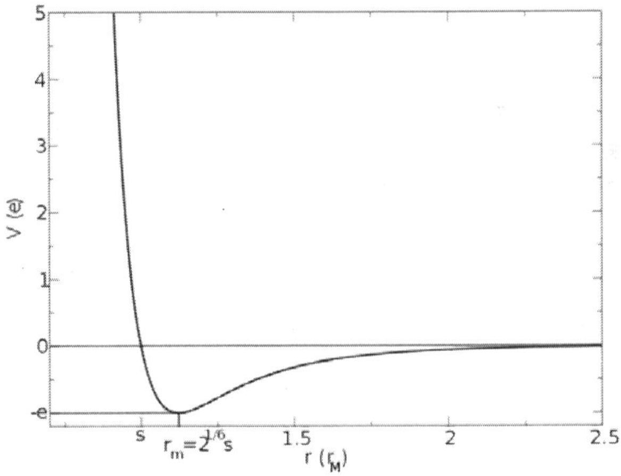

Figure: A graph of strength versus distance for the 12-6 Lennard-Jones potential.

The r^{-12} term, which is the repulsive term, describes Pauli repulsion at short ranges due to overlapping electron orbitals and the r^{-6} term, which is the attractive long-range term, describes attraction at long ranges (van der Waals force, or dispersion force). Whereas the functional form of the attractive term has a clear physical justification, the repulsive term has no theoretical justification. It is used because it approximates the Pauli repulsion well, and is more convenient due to the relative computational efficiency of calculating r^{12} as the square of r^6. The Lennard-Jones (12,6) potential is an approximation to the (exp-6) potential later proposed by R. A. Buckingham, in which the repulsive part is exponential:

$$V_B = \gamma\left[e^{-r/r_0} - \left(\frac{r_0}{r}\right)^6\right]$$

The L-J potential is a relatively good approximation and due to its simplicity is often used to describe the properties of gases, and to model dispersion and overlap interactions in molecular models. It is particularly accurate for noble gas atoms and is a good approximation at long and short distances for neutral atoms and molecules.

The lowest energy arrangement of an infinite number of atoms described by a Lennard-Jones potential is a hexagonal close-packing. On raising temperature, the lowest free energy arrangement becomes cubic close packing and then liquid. Under pressure the lowest energy structure switches between cubic and hexagonal close packing. Real materials include BCC structures as well.

Other more recent methods, such as the Stockmayer potential, describe the interaction of molecules more accurately. Quantum chemistry methods, Møller–Plesset perturbation theory, coupled cluster method or full configuration interaction can give extremely accurate results, but require large computational cost.

Alternative Expressions

There are many different ways of formulating the Lennard-Jones potential The following are some common forms.

AB Form

This form is a simplified formulation that is used by some simulation software packages:

$$V_{LJ}(r) = \frac{A}{r^{12}} - \frac{B}{r^6},$$

where $A = 4\varepsilon\sigma^{12}$ and $B = 4\varepsilon\sigma^6$. Conversely, $\sigma = \sqrt[6]{\frac{A}{B}}$ and $\varepsilon = \frac{B^2}{4A}$. This is the form in which Lennard-Jones wrote the 12-6 potential.

Truncated Lennard-Jones Potential

To save computational time, the Lennard-Jones potential is often truncated at a cut-off distance of $r_c = 2.5\sigma$, where

$$V_{LJ}(r_c) = V_{LJ}(2.5\sigma) = 4\varepsilon\left[\left(\frac{\sigma}{2.5\sigma}\right)^{12} - \left(\frac{\sigma}{2.5\sigma}\right)^{6}\right] \approx -0.0163\varepsilon \tag{1}$$

i.e., at $r_c = 2.5\sigma$, the Lennard-Jones potential, V_{LJ}, is about 1/60th of its minimum value, ε (the depth of the potential well). Beyond r_c, the truncated potential is set to zero.

For clarity, let V_{LJ} denote the LJ potential as defined above, i.e.,

$$V_{LJ}(r) = 4\varepsilon\left[\left(\frac{\sigma}{r}\right)^{12} - \left(\frac{\sigma}{r}\right)^{6}\right] \tag{2}$$

Then the truncated Lennard-Jones potential V_{LJ} is defined as follows

$$V_{LJ_{trunc}}(r)nn := \begin{cases} V_{LJ}(r) - V_{LJ}(r_c) & \text{for } r \le r_c \\ 0 & \text{for } r > r_c \end{cases} \tag{3}$$

It can be easily verified that $V_{\text{LJtrunc}}(r_c) = 0$, thus eliminating the jump discontinuity at $r = r_c$. Although the value of the (unshifted) Lennard Jones potential at $r = r_c = 2.5\sigma$ is rather small, the effect of the truncation can be significant, for instance on the gas–liquid critical point. Fortunately, the potential energy can be corrected for this effect in a mean field manner by adding so-called tail corrections.

Transition Dipole Moment

The Transition dipole moment or Transition moment, usually denoted $\mathbf{d}_{nm}$ for a transition between an initial state, m, and a final state, n, is the electric dipole moment associated with the transition between the two states. In general the transition dipole moment is a complex vector quantity that includes the phase factors associated with the two states. Its direction gives the polarization of the transition, which determines how the system will interact with an electromagnetic wave of a given polarization, while the square of the magnitude gives the strength of the interaction due to the distribution of charge within the system. The SI unit of the transition dipole moment is the Coulomb-metre (Cm); a more conveniently sized unit is the Debye (D).

The transition dipole moment for the $m \rightarrow n$ transition is given by the relevant off-diagonal element of the dipole matrix, which can be calculated from an integral taken over the product of the wavefunctions of the initial and final states of the transition, and the dipole moment operator,

$$\hat{\mathbf{d}} = e\left(\sum_i x_i, \sum_i y_i, \sum_i z_i\right),$$

where the summations are over the positions of the electrons in the system. Giving the transition dipole moment:

$$\mathbf{d}_{nm} = \int \Psi_n^* \, \hat{\mathbf{d}} \, \Psi_m \, d^3r = \langle \Psi_n | \hat{\mathbf{d}} | \Psi_m \rangle$$

where the integral is in principle over all space, but it can be restricted to the region in which the initial and final state wavefunctions are non-negligible.

Analogy with a Classical Dipole

A basic, phenomenological understanding of the transition dipole moment can be obtained by analogy with a classical dipole. While the comparison can be very useful, care must be taken to ensure that one does not fall into the trap of assuming they are the same.

In the case of two classical point charges, $+q$ and $-q$, with a displacement vector, r, pointing from the negative charge to the positive charge, the electric dipole moment is given by

$$\mathbf{p} = q\mathbf{r}\,.$$

In the presence of an electric field, such as that due to an electromagnetic wave, the two charges will experience a force in opposite directions, leading to a net torque on the dipole. The magnitude of the torque is proportional to both the magnitude of the charges and the separation between them, and varies with the relative angles of the field and the dipole:

$$|\tau| = |q\mathbf{r}|\,|\mathbf{E}|\sin\theta\,.$$

Similarly, the coupling between an electromagnetic wave and an atomic transition with transition dipole moment d_{nm}, depends on the charge distribution within the atom, the strength of the electric field, and the relative polarizations of the field and the transition. In addition, the transition dipole moment depends on the geometries and relative phases of the initial and final states.

Elementary Definition

In the simple case of two point charges, one with charge $+q$ and one with charge $-q$, the electric dipole moment p is:

$$\mathbf{p} = q\mathbf{d}$$

where d is the displacement vector pointing from the negative charge to the positive charge. Thus, the electric dipole moment vector p points from the negative charge to the positive charge. There is no inconsistency here, because the electric dipole moment has to do with *orientation* of the dipole, that is, the positions of the charges, and does not indicate the direction of the field originating in these charges.

An idealization of this two-charge system is the electrical point dipole consisting of two (infinite) charges only infinitesimally separated, but with a finite p.

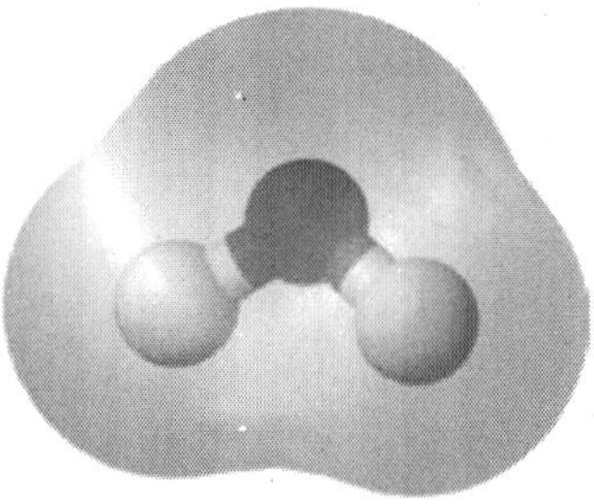

Figure: A molecule of water is polar because of the unequal sharing of its electrons in a "bent" structure. A separation of charge is present with negative charge in the middle (red shade), and positive charge at the ends (blue shade).

Torque

An object with an electric dipole moment is subject to a torque τ when placed in an external electric field. The torque tends to align the dipole with the field, and makes alignment an orientation of lower potential energy than misalignment. For a spatially uniform electric field E, the torque is given by:

$$\tau = \mathbf{p} \times \mathbf{E},$$

where p is the dipole moment, and the symbol "×" refers to the vector cross product. The field vector and the dipole vector define a plane, and the torque is directed normal to that plane with the direction given by the right-hand rule.

Expression (General Case)

More generally, for a continuous distribution of charge confined to a volume *V*, the corresponding expression for the dipole moment is:

$$\mathbf{p}(\mathbf{r}) = \int_V \rho(\mathbf{r}_0)(\mathbf{r}_0 - \mathbf{r})\, d^3\mathbf{r}_0,$$

where r locates the point of observation and d^3r_0 denotes an elementary volume in *V*. For an array of point charges, the charge density becomes a sum of Dirac delta functions:

$$\rho(\mathbf{r}) = \sum_{i=1}^{N} q_i \,\delta(\mathbf{r}-\mathbf{r}_i),$$

where each r_i is a vector from some reference point to the charge q_i. Substitution into the above integration formula provides:

$$\mathbf{p}(\mathbf{r}) = \sum_{i=1}^{N} q_i \int_V \delta(\mathbf{r}_0-\mathbf{r}_i)(\mathbf{r}_0-\mathbf{r})\, d^3\mathbf{r}_0 = \sum_{i=1}^{N} q_i(\mathbf{r}_i-\mathbf{r}).$$

This expression is equivalent to the previous expression in the case of charge neutrality and $N=2$. For two opposite charges, denoting the location of the positive charge of the pair as r_+ and the location of the negative charge as r_- :

$$\mathbf{p}(\mathbf{r})$$
$$= q_1(\mathbf{r}_1-\mathbf{r}) + q_2(\mathbf{r}_2-\mathbf{r}) = q(\mathbf{r}_+-\mathbf{r}) - q(\mathbf{r}_--\mathbf{r}) = q(\mathbf{r}_+-\mathbf{r}_-) = q\mathbf{d},$$

showing that the dipole moment vector is directed from the negative charge to the positive charge because the position vector of a point is directed outward from the origin to that point.

The dipole moment is most easily understood when the system has an overall neutral charge; for example, a pair of opposite charges, or a neutral conductor in a uniform electric field. For a system of charges with no net charge, visualized as an array of paired opposite charges, the relation for electric dipole moment is:

$$\begin{aligned}\mathbf{p}(\mathbf{r}) &= \sum_{i=1}^{N} \int_V q_i[\delta(\mathbf{r}_0-(\mathbf{r}_i+\mathbf{d}_i)) - \delta(\mathbf{r}_0-\mathbf{r}_i)](\mathbf{r}_0-\mathbf{r})\, d^3\mathbf{r}_0 \\ &= \sum_{i=1}^{N} q_i[\mathbf{r}_i+\mathbf{d}_i-\mathbf{r}-(\mathbf{r}_i-\mathbf{r})] \\ &= \sum_{i=1}^{N} q_i\mathbf{d}_i = \sum_{i=1}^{N} \mathbf{p}_i ,\end{aligned}$$

which is the vector sum of the individual dipole moments of the neutral charge pairs. (Because of overall charge neutrality, the dipole moment is independent of the observer's position r.) Thus, the value of p is independent of the choice of reference point, provided the overall charge of the system is zero.

When discussing the dipole moment of a non-neutral system, such as the dipole moment of the proton, a dependence on the choice of reference point arises. In such cases it is conventional to choose the reference point to be the centre of mass of the system, not some arbitrary origin. It might seem that the centre of charge is more reasonable reference point than the centre of mass, but it is clear that this results in a zero dipole moment. This convention ensures that the dipole moment is an intrinsic property of the system.

Potential and Field of an Electric Dipole

An ideal dipole consists of two opposite charges with infinitesimal separation. The potential and field of such an ideal dipole are found next as a limiting case of an example of two opposite charges at non-zero separation.

Two closely spaced opposite charges have a potential of the form:

$$\phi(\mathbf{r}) = \frac{q}{4\pi\varepsilon_0 \,|\mathbf{r} - \mathbf{r}_+|} - \frac{q}{4\pi\varepsilon_0 \,|\mathbf{r} - \mathbf{r}_-|},$$

with charge separation, d, defined as

$$\mathbf{d} = \mathbf{r}_+ - \mathbf{r}_-,$$

The radius to the centre of charge, R, and the unit vector in the direction of R are given by:

$$\mathbf{R} = \mathbf{r} - \frac{\mathbf{r}_+ + \mathbf{r}_-}{2}, \quad \widehat{\mathbf{R}} = \frac{\mathbf{R}}{R},$$

Taylor expansion in d/r allows this potential to be expressed as a series.

$$\phi(\mathbf{R}) = \frac{1}{4\pi\varepsilon_0}\frac{q\mathbf{d}\cdot\widehat{\mathbf{R}}}{R^2} + O\left(\frac{d^2}{R^2}\right) \approx \frac{1}{4\pi\varepsilon_0}\frac{\mathbf{p}\cdot\widehat{\mathbf{R}}}{R^2},$$

where higher order terms in the series are vanishing at large distances, R, compared to d. Here, the electric dipole moment p is, as above:

$$\mathbf{p} = q\mathbf{d}.$$

The result for the dipole potential also can be expressed as:

$$\phi(\mathbf{R}) = -\mathbf{p}\cdot\nabla\frac{1}{4\pi\varepsilon_0 R},$$

which relates the dipole potential to that of a point charge. A key point is that the potential of the dipole falls off faster with distance R than that of the point charge.

The electric field of the dipole is the negative gradient of the potential, leading to:

$$\mathbf{E} = \frac{3\mathbf{p}\cdot\widehat{\mathbf{R}}}{4\pi\varepsilon_0 R^3}\widehat{\mathbf{R}} - \frac{\mathbf{p}}{4\pi\varepsilon_0 R^3}\,.$$

Thus, although two closely spaced opposite charges are *not* an ideal electric dipole (because their potential at close approach is not that of a dipole), at distances much larger than their separation, their dipole moment p appears directly in their potential and field.

As the two charges are brought closer together (d is made smaller), the dipole term in the multipole expansion based on the ratio d/R becomes the only significant term at ever closer distances R, and in the limit of infinitesimal separation the dipole term in this expansion is all that matters. As d is made infinitesimal, however, the dipole charge must be made to increase to hold p constant. This limiting process results in a "point dipole".

Dipole Moment Density and Polarization Density

The dipole moment of an array of charges,

$$\mathbf{p} = \sum_{i=1}^{N} q_i \mathbf{d}_i\,,$$

determines the degree of polarity of the array, but for a neutral array it is simply a vector property of the array with no information about the array's absolute location. The dipole moment *density* of the array p(r) contains both the location of the array and its dipole moment. When it comes time to calculate the electric field in some region containing the array, Maxwell's equations are solved, and the information about the charge array is contained in the *polarization density* P(r) of Maxwell's equations. Depending upon how fine-grained an assessment of the electric field is required, more or less information about the charge array will have to be expressed by P(r). As explained below, sometimes it is sufficiently accurate to take P(r) = p(r). Sometimes a more detailed description is needed (for example, supplementing the dipole moment density with an additional quadrupole density) and sometimes even more elaborate versions of P(r) are necessary.

It now is explored just in what way the polarization density P(r) that enters Maxwell's equations is related to the dipole moment p of an overall neutral array of charges, and also to the dipole moment *density* p(r) (which describes not only the dipole moment, but also the array location). Only static situations are considered in what follows, so P has no time dependence, and there is no displacement current. First is some discussion of the polarization density P(r). That discussion is followed with several particular examples.

A formulation of Maxwell's equations based upon division of charges and currents into "free" and "bound" charges and currents leads to introduction of the D- and P-fields:

$$\mathbf{D} = \varepsilon_0 \mathbf{E} + \mathbf{P},$$

where P is called the polarization density. In this formulation, the divergence of this equation yields:

$$\nabla \cdot \mathbf{D} = \rho_f = \varepsilon_0 \nabla \cdot \mathbf{E} + \nabla \cdot \mathbf{P},$$

and as the divergence term in E is the *total* charge, and ρ_f is "free charge", we are left with the relation:

$$\nabla \cdot \mathbf{P} = -\rho_b,$$

with ρ_b as the bound charge, by which is meant the difference between the total and the free charge densities.

As an aside, in the absence of magnetic effects, Maxwell's equations specify that

$$\nabla \times \mathbf{E} = \mathbf{0},$$

which implies

$$\nabla \times (\mathbf{D} - \mathbf{P}) = \mathbf{0},$$

Applying Helmholtz decomposition:

$$(\mathbf{D} - \mathbf{P}) = -\nabla \varphi,$$

for some scalar potential φ, and:

$$\nabla \cdot (\mathbf{D} - \mathbf{P}) = \varepsilon_0 \nabla \cdot \mathbf{E} = \rho_f + \rho_b = -\nabla^2 \varphi.$$

Suppose the charges are divided into free and bound, and the potential is divided into

$$\varphi = \varphi_f + \varphi_b.$$

Satisfaction of the boundary conditions upon φ may be divided arbitrarily between φ_f and φ_b because only the sum φ must satisfy these conditions. It follows that P is simply proportional to the electric field due to the charges selected as bound, with boundary conditions that prove convenient. In particular, when *no* free charge is present, one possible choice is P = ε_0 E.

Next is discussed how several different dipole-moment descriptions of a medium relate to the polarization entering Maxwell's equations.

Medium with Charge and Dipole Densities

As described next, a model for polarization moment density p(r) results in a polarization

$$\mathbf{P}(\mathbf{r}) = \mathbf{p}(\mathbf{r})$$

restricted to the same model. For a smoothly varying dipole moment distribution p(r), the corresponding bound charge density is simply

$$\nabla \cdot \mathbf{p}(\mathbf{r}) = \rho_b \ .$$

However, in the case of a p(r) that exhibits an abrupt step in dipole moment at a boundary between two regions, $\nabla \cdot \mathrm{p}(\mathrm{r})$ exhibits a surface charge component of bound charge. This surface charge can be treated through a surface integral, or by using discontinuity conditions at the boundary, as illustrated in the various examples below.

As a first example relating dipole moment to polarization, consider a medium made up of a continuous charge density ρ(r) and a continuous dipole moment distribution p(r). The potential at a position r is:

$$\phi(\mathbf{r}) = \frac{1}{4\pi\varepsilon_0}\int \frac{\rho(\mathbf{r}_0)}{|\,\mathbf{r}-\mathbf{r}_0\,|} d^3\mathbf{r}_0 + \frac{1}{4\pi\varepsilon_0}\int \frac{\mathbf{p}(\mathbf{r}_0)\cdot(\mathbf{r}-\mathbf{r}_0)}{|\,\mathbf{r}-\mathbf{r}_0\,|^3} d^3\mathbf{r}_0,$$

where ρ(r) is the unpaired charge density, and p(r) is the dipole moment density. Using an identity:

$$\nabla_{\mathbf{r}_0} \frac{1}{|\,\mathbf{r}-\mathbf{r}_0\,|} = \frac{\mathbf{r}-\mathbf{r}_0}{|\,\mathbf{r}-\mathbf{r}_0\,|^3}$$

the polarization integral can be transformed:

$$\frac{1}{4\pi\varepsilon_0}\int \frac{\mathbf{p}(\mathbf{r}_0)\cdot(\mathbf{r}-\mathbf{r}_0)}{|\,\mathbf{r}-\mathbf{r}_0\,|^3} d^3\mathbf{r}_0 = \frac{1}{4\pi\varepsilon_0}\int \mathbf{p}(\mathbf{r}_0)\cdot\nabla_{\mathbf{r}_0} \frac{1}{|\,\mathbf{r}-\mathbf{r}_0\,|} d^3\mathbf{r}_0,$$

$$= \frac{1}{4\pi\varepsilon_0}\int \nabla_{\mathbf{r}_0}\cdot\left(\mathbf{p}(\mathbf{r}_0)\frac{1}{|\,\mathbf{r}-\mathbf{r}_0\,|}\right) d^3\mathbf{r}_0 - \frac{1}{4\pi\varepsilon_0}\int \frac{\nabla_{\mathbf{r}_0}\cdot\mathbf{p}(\mathbf{r}_0)}{|\,\mathbf{r}-\mathbf{r}_0\,|} d^3\mathbf{r}_0,$$

The first term can be transformed to an integral over the surface bounding the volume of integration, and contributes a surface charge density, discussed later. Putting this result back into the potential, and ignoring the surface charge for now:

$$\phi(\mathbf{r}) = \frac{1}{4\pi\varepsilon_0}\int \frac{\rho(\mathbf{r}_0) - \nabla_{\mathbf{r}_0}\cdot\mathbf{p}(\mathbf{r}_0)}{|\mathbf{r}-\mathbf{r}_0|} d^3\mathbf{r}_0 ,$$

where the volume integration extends only up to the bounding surface, and does not include this surface. The potential is determined by the total charge, which the above shows consists of:

$$\rho_{\text{total}}(\mathbf{r}_0) = \rho(\mathbf{r}_0) - \nabla_{\mathbf{r}_0}\cdot\mathbf{p}(\mathbf{r}_0) ,$$

showing that:

$$-\nabla_{\mathbf{r}_0}\cdot\mathbf{p}(\mathbf{r}_0) = \rho_b .$$

In short, the dipole moment density p(r) plays the role of the polarization density P for this medium. Notice, p(r) has a non-zero divergence equal to the bound charge density (as modelled in this approximation). It may be noted that this approach can be extended to include all the multipoles: dipole, quadrupole, etc. Using the relation:

$$\nabla\cdot\mathbf{D} = \rho_f ,$$

the polarization density is found to be:

$$\mathbf{P}(\mathbf{r}) = \mathbf{p}_{\text{dip}} - \nabla\cdot\mathbf{p}_{\text{quad}} + \ldots ,$$

where the added terms are meant to indicate contributions from higher multipoles. Evidently, inclusion of higher multipoles signifies that the polarization density P no longer is determined by a dipole moment density p. For example, in considering scattering from a charge array, different multipoles scatter an electromagnetic wave differently and independently, requiring a representation of the charges that goes beyond the dipole approximation.

Surface Charge

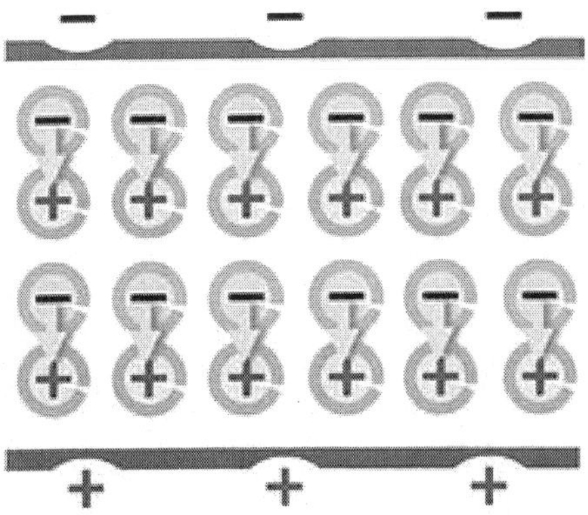

Figure: A uniform array of identical dipoles is equivalent to a surface charge.

Above, discussion was deferred for the leading divergence term in the expression for the potential due to the dipoles. This term results in a surface charge. The figure shows a uniform array of identical dipoles between two surfaces. Internally, the heads and tails of dipoles are adjacent and cancel. At the bounding surfaces, however, no cancellation occurs. Instead, on one surface the dipole heads create a positive surface charge, while at the opposite surface the dipole tails create a negative surface charge. These two opposite surface charges create a net electric field in a direction opposite to the direction of the dipoles.

This idea is given mathematical form using the potential expression above. The potential is:

$$\phi(\mathbf{r}) = \frac{1}{4\pi\varepsilon_0}\int \nabla_{\mathbf{r}_0}\cdot\left(\mathbf{p}(\mathbf{r}_0)\frac{1}{|\mathbf{r}-\mathbf{r}_0|}\right)d^3\mathbf{r}_0 - \frac{1}{4\pi\varepsilon_0}\int\frac{\nabla_{\mathbf{r}_0}\cdot\mathbf{p}(\mathbf{r}_0)}{|\mathbf{r}-\mathbf{r}_0|}d^3\mathbf{r}_0\,.$$

Using the divergence theorem, the divergence term transforms into the surface integral:

$$\frac{1}{4\pi\varepsilon_0}\int \nabla_{\mathbf{r}_0}\cdot\left(\mathbf{p}(\mathbf{r}_0)\frac{1}{|\mathbf{r}-\mathbf{r}_0|}\right)d^3\mathbf{r}_0$$

$$= \frac{1}{4\pi\varepsilon_0}\int\frac{\mathbf{p}(\mathbf{r}_0)\cdot d\mathbf{A}_0}{|\mathbf{r}-\mathbf{r}_0|}\,,$$

with dA_0 an element of surface area of the volume. In the event that p(r) is a constant, only the surface term survives:

$$\phi(\mathbf{r}) = \frac{1}{4\pi\varepsilon_0}\int\frac{1}{|\mathbf{r}-\mathbf{r}_0|}\,\mathbf{p}\cdot d\mathbf{A}_0\,,$$

with dA_0 an elementary area of the surface bounding the charges. In words, the potential due to a constant p inside the surface is equivalent to that of a *surface charge*

$$\sigma = \mathbf{p}\cdot d\mathbf{A}$$

which is positive for surface elements with a component in the direction of p and negative for surface elements pointed oppositely. (Usually the direction of a surface element is taken to be that of the outward normal to the surface at the location of the element.)

If the bounding surface is a sphere, and the point of observation is at the centre of this sphere, the integration over the surface of the sphere is zero: the positive and negative surface charge contributions to the potential cancel. If the point of observation is off-centre, however, a net potential can result (depending upon the situation) because the positive and negative charges are at different distances from the point of observation. The field due to the surface charge is:

$$\mathbf{E}(\mathbf{r}) = -\frac{1}{4\pi\varepsilon_0}\nabla_r \int \frac{1}{|\mathbf{r}-\mathbf{r}_0|}\,\mathbf{p}\cdot d\mathbf{A}_0 ,$$

which, at the centre of a spherical bounding surface is not zero (the *fields* of negative and positive charges on opposite sides of the centre add because both fields point the same way) but is instead:

$$\mathbf{E} = -\frac{\mathbf{p}}{3\varepsilon_0} .$$

If we suppose the polarization of the dipoles was induced by an external field, the polarization field opposes the applied field and sometimes is called a *depolarization field*. In the case when the polarization is *outside* a spherical cavity, the field in the cavity due to the surrounding dipoles is in the *same* direction as the polarization.

In particular, if the electric susceptibility is introduced through the approximation:

$$\mathbf{p}(\mathbf{r}) = \varepsilon_0 \chi(\mathbf{r})\mathbf{E}(\mathbf{r}) ,$$

then:

$$\nabla\cdot\mathbf{p}(\mathbf{r}) = \nabla\cdot\left(\chi(\mathbf{r})\varepsilon_0\mathbf{E}(\mathbf{r})\right) = -\rho_b .$$

Whenever $\chi(\mathrm{r})$ is used to model a step discontinuity at the boundary between two regions, the step produces a surface charge layer. For example, integrating along a normal to the bounding surface from a point just interior to one surface to another point just exterior:

$$\varepsilon_0\hat{\mathbf{n}}\cdot\left[\chi(\mathbf{r}_+)\mathbf{E}(\mathbf{r}_+) - \chi(\mathbf{r}_-)\mathbf{E}(\mathbf{r}_-)\right] = \frac{1}{A_n}\int d\Omega_n\ \rho_b = 0 ,$$

where A_n, Ω_n indicate the area and volume of an elementary region straddling the boundary between the regions, and $\hat{\mathbf{n}}$ a unit normal to the surface. The right side vanishes as the volume shrinks, inasmuch as ρ_b is finite, indicating a discontinuity in E, and therefore a surface charge. That is, where the modelled medium includes a step in permittivity, the polarization density corresponding to the dipole moment density

$$\mathbf{p}(\mathbf{r}) = \chi(\mathbf{r})\mathbf{E}(\mathbf{r})$$

necessarily includes the contribution of a surface charge.

It may be noted that a physically more realistic modelling of p(r) would cause the dipole moment density to taper off continuously to zero at the boundary of the confining region, rather than making a sudden step to zero density. Then the surface charge becomes zero at the boundary, and the surface charge is replaced by the divergence of a continuously varying dipole-moment density.

Dielectric Sphere in Uniform External Electric Field

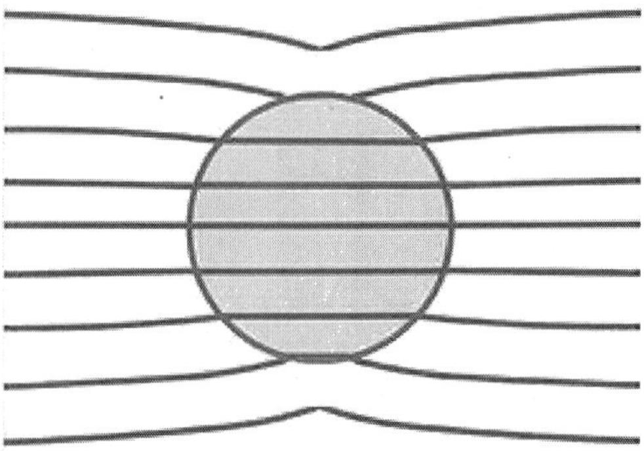

Figure: Field lines of the D-field in a dielectric sphere with greater susceptibility than its surroundings, placed in a previously-uniform field. The field lines of the E-field are not shown: These point in the same directions, but many field lines start and end on the surface of the sphere, where there is bound charge. As a result, the density of E-field lines is lower inside the sphere than outside, which corresponds to the fact that the E-field is weaker inside the sphere than outside.

The above general remarks about surface charge are made more concrete by considering the example of a dielectric sphere in a uniform electric field. The sphere is found to adopt a surface charge related to the dipole moment of its interior.

A uniform external electric field is supposed to point in the z-direction, and spherical-polar coordinates are introduced so the potential created by this field is:

$$\phi_\infty = -E_\infty z = -E_\infty r \cos\theta \, .$$

The sphere is assumed to be described by a dielectric constant κ, that is,

$$\mathbf{D} = \kappa \epsilon_0 \mathbf{E} \, ,$$

and inside the sphere the potential satisfies Laplace's equation. Skipping a few details, the solution inside the sphere is:

$$\phi = Ar \cos\theta \, ,$$

while outside the sphere:

$$\phi_{>} = \left(Br + \frac{C}{r^2} \right) \cos\theta .$$

At large distances, $\varphi_{>} \rightarrow \varphi_{\_}$ so $B = -E_{\_}$. Continuity of potential and of the radial component of displacement $D = \kappa\varepsilon_0 E$ determine the other two constants. Supposing the radius of the sphere is R,

$$A = -\frac{3}{\kappa + 2} E_{\infty} ; C = \frac{\kappa - 1}{\kappa + 2} E_{\infty} R^3 ,$$

As a consequence, the potential is:

$$\phi_{>} = \left(-r + \frac{\kappa - 1}{\kappa + 2} \frac{R^3}{r^2} \right) E_{\infty} \cos\theta ,$$

which is the potential due to applied field and, in addition, a dipole in the direction of the applied field (the z-direction) of dipole moment:

$$\mathbf{p} = 4\pi\varepsilon_0 \left(\frac{\kappa - 1}{\kappa + 2} R^3 \right) \mathbf{E}_{\infty} ,$$

or, per unit volume:

$$\frac{\mathbf{p}}{V} = 3\varepsilon_0 \left(\frac{\kappa - 1}{\kappa + 2} \right) \mathbf{E}_{\infty} .$$

The factor $(\kappa-1)/(\kappa+2)$ is called the Clausius-Mossotti factor and shows that the induced polarization flips sign if $\kappa < 1$. Of course, this cannot happen in this example, but in an example with two different dielectrics κ is replaced by the ratio of the inner to outer region dielectric constants, which can be greater or smaller than one. The potential inside the sphere is:

$$\phi_{<} = -\frac{3}{\kappa + 2} E_{\infty} r \cos\theta ,$$

leading to the field inside the sphere:

$$-\nabla\phi_{<} = \frac{3}{\kappa + 2} \mathbf{E}_{\infty} = \left(1 - \frac{\kappa - 1}{\kappa + 2} \right) \mathbf{E}_{\infty} ,$$

showing the depolarizing effect of the dipole. Notice that the field inside the sphere is *uniform* and parallel to the applied field. The dipole moment is uniform throughout the interior of the sphere. The surface charge density on the sphere is the difference between the radial field components:

$$\sigma = 3\varepsilon_0 \frac{\kappa - 1}{\kappa + 2} E_\infty \cos\theta = \frac{1}{V} \mathbf{p}\cdot\hat{\mathbf{R}} .$$

This linear dielectric example shows that the dielectric constant treatment is equivalent to the uniform dipole-moment model and leads to zero charge everywhere except for the surface charge at the boundary of the sphere.

General Media

If observation is confined to regions sufficiently remote from a system of charges, a multipole expansion of the exact polarization density can be made.

By truncating this expansion (for example, retaining only the dipole terms, or only the dipole and quadrupole terms, or *etc.*), the results of the previous section are regained.

In particular, truncating the expansion at the dipole term, the result is indistinguishable from the polarization density generated by a uniform dipole moment confined to the charge region. To the accuracy of this dipole approximation, as shown in the previous section, the dipole moment *density* p(r) (which includes not only p but the location of p) serves as P(r).

At locations *inside* the charge array, to connect an array of paired charges to an approximation involving only a dipole moment density p(r) requires additional considerations.

The simplest approximation is to replace the charge array with a model of ideal (infinitesimally spaced) dipoles. In particular, as in the example above that uses a constant dipole moment density confined to a finite region, a surface charge and depolarization field results. A more general version of this model (which allows the polarization to vary with position) is the customary approach using a electric susceptibility or electrical permittivity.

A more complex model of the point charge array introduces an effective medium by averaging the microscopic charges; for example, the averaging can arrange that only dipole fields play a role.

A related approach is to divide the charges into those nearby the point of observation, and those far enough away to allow a multipole expansion. The nearby charges then give rise to *local field effects*. In a common model of this type, the distant charges are treated as a homogeneous medium using a dielectric constant, and the nearby charges are treated only in a dipole approximation. The approximation of a medium or an array of charges by only dipoles and their associated dipole moment density is sometimes called the *point dipole* approximation, the *discrete dipole approximation*, or simply the *dipole approximation*.

Dipole Moments of Fundamental Particles

Much experimental work is continuing on measuring the electric dipole moments (EDM) of fundamental and composite particles, namely those of the neutron and electron.

As EDMs violate both the Parity (P) and Time (T) symmetries, their values yield a mostly model-independent measure (assuming CPT symmetry is valid) of CP-violation in nature. Therefore, values for these EDMs place strong constraints upon the scale of CP-violation that extensions to the standard model of particle physics may allow.

Indeed, many theories are inconsistent with the current limits and have effectively been ruled out, and established theory permits a much larger value than these limits, leading to the strong CP problem and prompting searches for new particles such as the axion.

Current generations of experiments are designed to be sensitive to the supersymmetry range of EDMs, providing complementary experiments to those done at the LHC.

Dipole moments of Molecules

Dipole moments in molecules are responsible for the behaviour of a substance in the presence of external fields. The dipoles tend to be aligned to the external field which can be constant or time-dependent. This effect forms the basis of a modern experimental technique called Dielectric spectroscopy.

Dipole moments can be found in common molecules such as water and also in biomolecules such as proteins.

By means of the total dipole moment of some material one can compute the dielectric constant which is related to the more intuitive concept of conductivity. If $\mathcal{M}_{\text{Tot}}$ is the total dipole moment of the sample, then the dielectric constant is given by,

$$\epsilon = 1 + k\langle \mathcal{M}_{\text{Tot}}^2 \rangle$$

where k is a constant and

$\langle \mathcal{M}_{\text{Tot}}^2 \rangle = \langle \mathcal{M}_{\text{Tot}}(t=0)\mathcal{M}_{\text{Tot}}(t=0)\rangle$ is the time correlation function of the total dipole moment. In general the total dipole moment have contributions coming from translations and rotations of the molecules in the sample,

$$\mathcal{M}_{\text{Tot}} = \mathcal{M}_{\text{Trans}} + \mathcal{M}_{\text{Rot}}.$$

Therefore, the dielectric constant (and the conductivity) has contributions from both terms. This approach can be generalized to compute the frequency dependent dielectric function.

When an atom or molecule interacts with an electromagnetic wave of frequency ω, it can undergo a transition from an initial to a final state of energy difference $\hbar\omega$ through the coupling of the electromagnetic field to the transition dipole moment.

When this transition is from a lower energy state to a higher energy state, this results in the absorption of a photon. A transition from a higher energy state to a lower energy state, results in the emission of a photon. If the charge, e, is omitted from the electric dipole operator during this calculation, one obtains R_a as used in oscillator strength.

The transition dipole moment is useful for determining if transitions are allowed under the electric dipole interaction.

For example, the transition from a bonding π orbital to an antibonding π^* orbital is allowed because the integral defining the transition dipole moment is nonzero.

Such a transition occurs between an even and an odd orbital; the dipole operator is an odd function of r, hence the integrand is an even function. The integral of an odd function over symmetric limits returns a value of zero, while for an even function this is not *necessarily* the case. This result is reflected in the parity selection rule for electric dipole transitions.

Vibrational spectroscopy provides the most definitive means of identifying the surface species generated upon molecular adsorption and the species generated by surface reactions.

In principle, any technique that can be used to obtain vibrational data from solid state or gas phase samples (IR, Raman etc.) can be applied to the study of surfaces - in addition there are a number of techniques which have been specifically developed to study the vibrations of molecules at interfaces (EELS, SFG etc.). There are, however, only two techniques that are routinely used for vibrational studies of molecules on surfaces- these are :

1. IR Spectroscopy (of various forms, e.g. RAIRS, MIR)
2. Electron Energy Loss Spectroscopy (EELS)

IR Spectroscopy

There are a number of ways in which the IR technique may be implemented for the study of adsorbates on surfaces.

For solid samples possessing a high surface area:

Transmission IR Spectroscopy: employing the same basic experimental geometry as that used for liquid samples and mulls.

This is often used for studies on supported metal catalysts where the large metallic surface area permits a high concentration of adsorbed species to be sampled.

The solid sample must, of course, be IR transparent over an appreciable wavelength range.

Diffuse Reflectance IR Spectroscopy (DRIFTS): in which the diffusely scattered IR radiation from a sample is collected, refocused and analysed. This modification of the IR technique can be employed with high surface area catalytic samples that are not sufficiently transparent to be studied in transmission.

For Studies on Low Surface Area Samples (e.g. Single Crystals): Reflection-Absorption IR Spectroscopy (RAIRS): where the IR beam is specularly reflected from the front face of a highly-reflective sample, such as a metal single crystal surface.

Multiple Internal Reflection Spectroscopy (MIR) : in which the IR beam is passed through a thin, IR transmitting sample in a manner such that it alternately undergoes total internal reflection from the front and rear faces of the sample. At each reflection, some of the IR radiation may be absorbed by species adsorbed on the solid surface - hence the alternative name of Attenuated Total Reflection (ATR).

RAIRS - the Study of Adsorbates on Metallic Surfaces by Reflection IR Spectroscopy

It can be shown theoretically that the best sensitivity for IR measurements on metallic surfaces is obtained using a grazing-incidence reflection of the IR light.

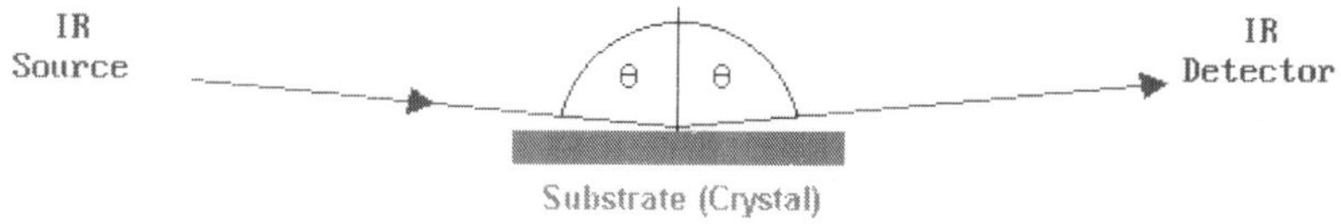

Figure: Furthermore, since it is an optical (photon in/photon out) technique it is not necessary for such studies to be carried out in vacuum. The technique is not inherently surface-specific, but

- there is no bulk signal to worry about
- the surface signal is readily distinguishable from gas-phase absorptions using polarization effects.

One major problem, is that of sensitivity (i.e. the signal is usually very weak owing to the small number of adsorbing molecules). Typically, the sampled area is ca. 1 cm^2 with less than 10^{15} adsorbed molecules (i.e. about 1 nanomole). With modern FTIR spectrometres, however, such small signals (0.01% - 2% absorption) can still be recorded at relatively high resolution (ca. 1 cm^{-1}). For a number of practical reasons, low frequency modes (< 600 cm^{-1}) are not generally observable - this means that it is not usually possible to see the vibration of the metal-adsorbate bond and attention is instead concentrated on the intrinsic vibrations of the adsorbate species in the range 600 - 3600 cm^{-1}.

Selection Rules

The observation of vibrational modes of adsorbates on metallic substrates is subject to the *surface dipole selection rule.*

This states that only those vibrational modes which give rise to an oscillating dipole perpendicular (normal) to the surface are IR active and give rise to an observable absorption band.

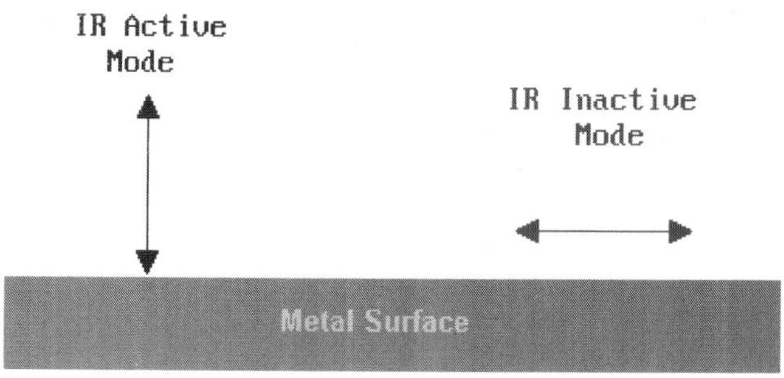

Figure: Further information on the selection rules for surface IR spectroscopy can be found in the review by Sheppard & Erkelens. It also needs to be remembered that even if a transition is allowed it may still be very weak if the transition moment is small.

Examples of Reflection IR Spectra

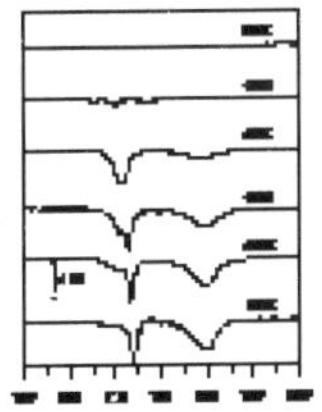

Figure: Nitric oxide (NO) adsorption on a Pt surface.

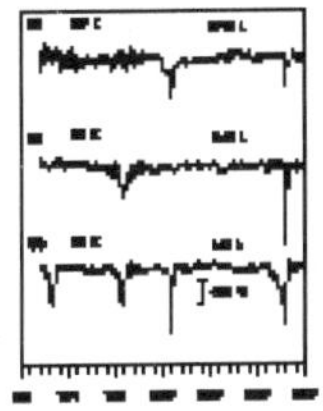

Figure: HCN adsorption on a Pt surface

Electron Energy Loss Spectroscopy (EELS)

This is a technique utilising the inelastic scattering of low energy electrons in order to measure vibrational spectra of surface species: superficially, it can be considered as the electron-analogue of Raman spectroscopy. To avoid confusion with other electron energy loss techniques it is sometimes referred to as

HREELS - high resolution EELS *or* VELS - vibrational ELS

Since the technique employs low energy electrons, it is necessarily restricted to use in high vacuum (HV) and UHV environments - however, the use of such low energy electrons ensures that it is a surface specific technique and, arguably, it is the vibrational technique of choice for the study of most adsorbates on single crystal substrates. The basic experimental geometry is fairly simple as illustrated schematically below - it involves using an electron monochromator to give a well-defined beam of electrons of a fixed incident energy, and then analysing the scattered electrons using an appropriate electron energy analyser.

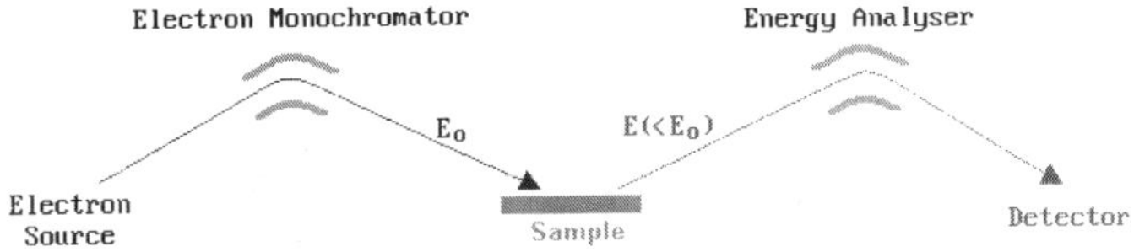

Figure: A substantial number of electrons are elastically scattered ($E = E_0$)- this gives rise to a strong *elastic peak* in the spectrum.

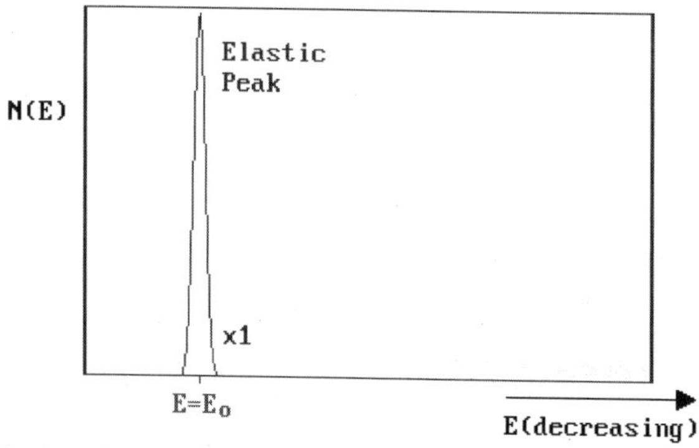

Figure: On the low kinetic energy side of this main peak ($E < E_0$), additional weak peaks are superimposed on a mildly sloping background. These peaks correspond to electrons which have undergone discrete energy losses during the scattering from the surface.

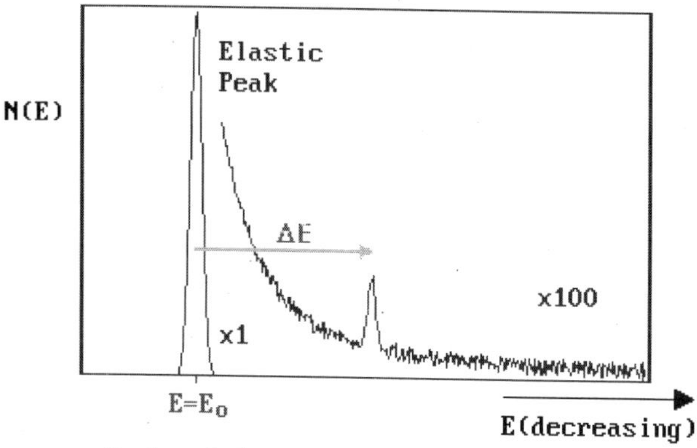

Figure: The magnitude of the energy loss, $\Delta E = (E_0 - E)$, is equal to the vibrational quantum (i.e. the energy) of the vibrational mode of the adsorbate excited in the inelastic scattering process. In practice, the incident energy (E_0) is usually in the range 5-10 eV (although occasionally up to 200 eV) and the data is normally plotted against the energy loss (frequently measured in meV).

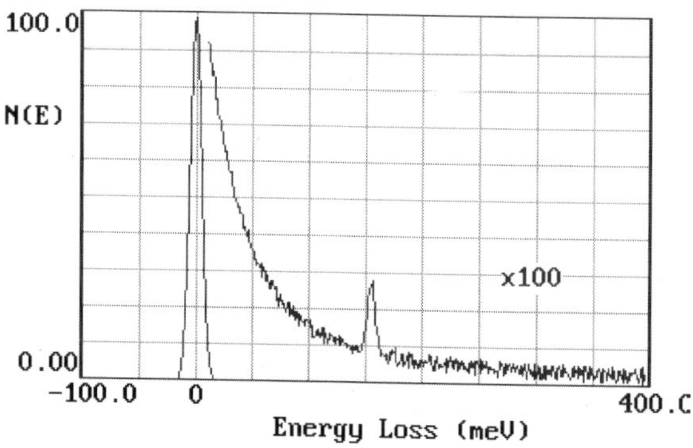

Selection Rules

The selection rules that determine whether a vibrational band may be observed depend upon the nature of the substrate and also the experimental geometry: specifically the angles of the incident and (analysed) scattered beams with respect to the surface.

For metallic substrates and a specular geometry, scattering is principally by a long-range dipole mechanism.

In this case the loss features are relatively intense, but only those vibrations giving rise to a dipole change normal to the surface can be observed.

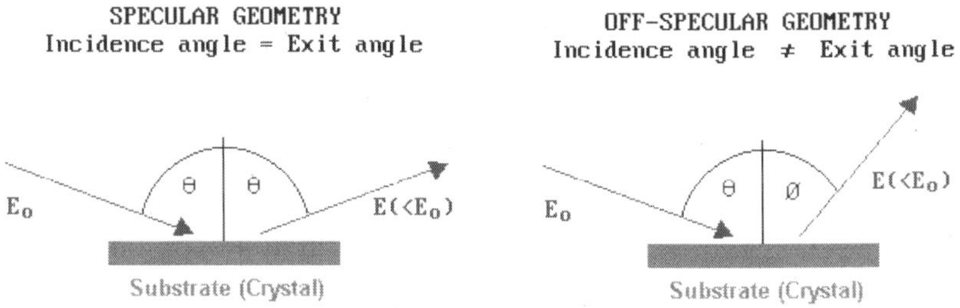

By contrast, in an off-specular geometry, electrons lose energy to surface species by a short-range impact scattering mechanism. In this case the loss features are relatively weak but all vibrations are allowed and may be observed.

If spectra can be recorded in *both* specular and off-specular modes the selection rules for metallic substrates can be put to good use - helping the investigator to obtain more definitive identification of the nature and geometry of the adsorbate species.

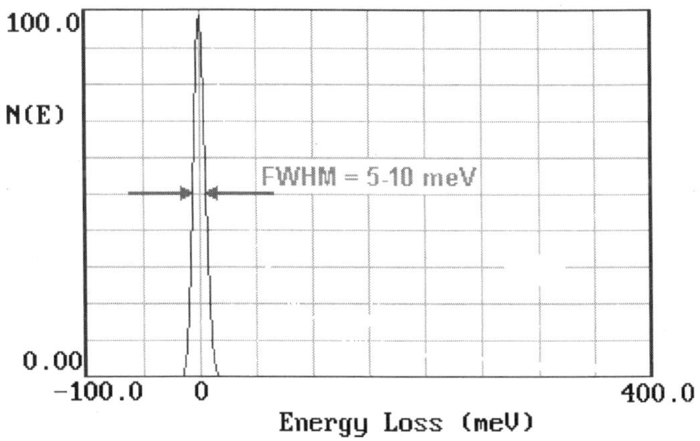

The resolution of the technique (despite the HREELS acronym !) is generally rather poor ; 40-80 cm^{-1} is not untypical. A measure of the instrumental resolution is given by looking at the FWHM (full-width at half maximum) of the elastic peak.

This poor resolution can cause problems in distinguishing between closely similar surface species - however, recent improvements in instrumentation have opened up the possibility of much better spectral resolution (< 10 cm^{-1}) and will undoubtedly enhance the utility of the technique.

In summary, there are both advantages and disadvantages in utilising EELS, as opposed to IR techniques, for the study of surface species It offers the advantages of ...

- high sensitivity
- variable selection rules
- spectral acquisition to below 400 cm^{-1}

but suffers from the limitations of ...

- use of low energy electrons (requiring a HV environment and hence the need for low temperatures to study weakly-bound species, and also the use of magnetic shielding to reduce the magnetic field in the region of the sample)
- requirement for flat, preferably conducting, substrates
- lower resolution

Applications of Vibrational Spectroscopy

One of the classic examples of an area in which vibrational spectroscopy has contributed significantly to the understanding of the surface chemistry of an adsorbate is that of:

Molecular Adsorption of CO on Metallic Surfaces

Adsorbed carbon monoxide usually gives rise to strong absorptions in both the IR and EELS spectra at the (C-O) stretching frequency. The metal-carbon stretching mode (ca. 400 $cm^{-1)}$ is usually also accessible to EELS.

The interpretation of spectra of CO as an adsorbed surface species draws heavily upon IR spectra from related inorganic cluster and coordination complexes - the structure of such complexes usually being available from x-ray single crystal diffraction measurements. This comparison suggests that the CO stretching frequency can provide a good indication of the surface coordination of the molecule: as a rough guideline,

	ν(C-O)
CO (gas phase)	2143 cm^{-1}
Terminal CO	2100 - 1920 cm^{-1}
Bridging (2f site)	1920 - 1800 cm^{-1}
Bridging (3f / 4f site)	< 1800 cm^{-1}

Example: RAIR spectrum for CO chemisorbed on a clean Pt surface.

The reduction in the stretching frequency of terminally-bound CO from the value observed for the gas phase molecule (2143 cm^{-1}) can be explained in terms of the Dewar-Chatt or *Blyholder model* for the bonding of CO to metals.

Chapter 5

Resonance Raman Spectroscopy

Resonance Raman (RR) spectroscopy is a specialised implementation of the more general Raman spectroscopy. As in Raman spectroscopy, RR spectroscopy provides information about the vibrations of molecules, and can also be used for identifying unknown substances. RR spectroscopy has found wide application to the analysis of bioinorganic molecules. Although the technique uses the same part of the electromagnetic spectrum as infrared (IR) spectroscopy, the two methods are actually complementary. Both are used to measure the energy required to change the vibrational state of a chemical compound.

IR spectroscopy involves measuring the direct absorption of photons with the appropriate energy to excite molecular bond vibrations. The wavelengths of these photons lie in the infrared region of the spectrum, hence the name of the technique. Raman spectroscopy measures the excitation of bond vibrations in an indirect manner. The two methods are complementary because some vibrational transitions that are observed in IR spectroscopy are not observed in Raman spectroscopy, and vice versa. RR spectroscopy is an improvement of traditional Raman spectroscopy that has increased sensitivity and is better suited for the study of complicated systems.

X-Ray Raman Scattering

In the x-ray region, enough energy is available for making electronic transitions possible. At core level resonances, X-Ray Raman Scattering can become the dominating part of the x-ray fluorescence spectrum. This is due to the resonant behaviour of the Kramers-Heisenberg formula in which the denominator is minimised for incident energies that equal a core level. This type of scattering is also known as Resonant inelastic X-ray scattering (RIXS). In the soft x-ray range, RIXS has been shown to reflect crystal field excitations, which are often hard to observe with any other technique. Application of RIXS to strongly correlated materials is of particular value for gaining knowledge about their electronic structure. For certain wide band materials such as graphite, RIXS has been shown to (nearly) conserve crystal momentum and thus has found use as a complementary bandmapping technique.

Raman spectroscopy utilises the phenomenon of scattering. When light passes through a transparent sample, a fraction of the light is scattered in all directions. Most of the scattered photons are of the same wavelength of the incident light. This is known as Rayleigh scattering. However, in 1928, physicists C.V. Raman and K.S. Krishnan, and independently Grigory Landsberg and Leonid Mandelstam discovered that a small fraction of the scattered light had a different wavelength. Furthermore, this difference depended on the molecules present in the sample. For these observations, and his explanation of the phenomenon, Raman was awarded the 1930 Nobel Prize in physics. His explanation is now known as the theory of Raman scattering.

In the years following its discovery, Raman spectroscopy was used to provide the first catalog of molecular vibrational frequencies. Originally, heroic measures were required to obtain Raman spectra due to the low sensitivity of the technique. Typically, the sample was held in a long tube and illuminated along its length with a beam of filtered monochromatic light generated by a gas discharge lamp. The photons that were scattered by the sample were collected through an optical flat at the end of the tube. To maximise the sensitivity, the sample was highly concentrated (1 M or more) and relatively large volumes (5 mL or more) were used. Consequently, the use of Raman spectroscopy dwindled when commercial IR spectrophotometres became available in the 1940s. However, the advent of the laser in the 1960s resulted in simplified Raman spectroscopy instruments and also boosted the sensitivity of the technique. This has revived the use of Raman spectroscopy as a common analytical technique.

Theory of Raman Scattering

Degrees of Freedom

For any given chemical compound, there are a total of 3N degrees of freedom, where N is the number of atoms in the compound. This number arises from the ability of each atom in a molecule to move in three different directions (x, y, and z). When dealing with molecules, it is more common to consider the movement of the molecule as a whole. Consequently, the 3N degrees of freedom are partitioned into molecular translational, rotational, and vibrational motion. Three of the degrees of freedom correspond to translational motion of the molecule as a whole (along each of the three spatial dimensions). Similarly, three degrees of freedom correspond to rotations of the molecule about the x, y, and z-axes. However, linear molecules only have two rotations because rotations along the bond axis do not change the positions of the atoms in the molecule. The remaining degrees of freedom correspond to molecular vibrational modes. These modes include stretching and bending motions of the chemical bonds of the molecule. For a linear molecule, the number of vibrational modes is:

$$3N - 3 - 2 = 3N - 5$$

whereas for a non-linear molecule the number of vibrational modes are

$$3N - 6$$

Molecular Vibrations and Infrared Radiation

The frequencies of molecular vibrations range from less than 10^{12} to approximately 10^{14} Hz. These frequencies correspond to radiation in the infrared (IR) region of the electromagnetic spectrum. At any given instant, each molecule in a sample has a certain amount of vibrational energy. However, the amount of vibrational energy that a molecule has continually changes due to collisions and other interactions with other molecules in the sample.

At room temperature, most of molecules will be in the lowest energy state, which is known as the ground state. A few molecules will be in higher energy states, which are known as excited states. The fraction of molecules occupying a given vibrational mode at a given temperature can be calculated using the Boltzmann distribution. Performing such a calculation shows that, for relatively low temperatures (such as those used for most routine spectroscopy), most of the molecules occupy the ground vibrational state. Such a molecule can be excited to a higher vibrational mode through the direct absorption of a photon of the appropriate energy. This is the mechanism by which IR spectroscopy operates: infrared radiation is passed through the sample, and the intensity of the transmitted light is compared with that of the incident light. A reduction in intensity at a given wavelength of light indicates the absorption of energy by a vibrational transition. The energy, E, of a photon is

$$E = h\nu$$

where h is Planck's constant and υ is the frequency of the radiation. Thus, the energy required for such transition may be calculated if the frequency of the incident radiation is known.

Raman Scattering

It is also possible to observe molecular vibrations by an inelastic scattering process. In inelastic scattering, an absorbed photon is re-emitted with lower energy. In Raman scattering, the difference in energy between the absorbed and re-emitted photons corresponds to the energy required to excite a molecule to a higher vibrational mode.

Typically, in Raman spectroscopy high intensity laser radiation with wavelengths in either the visible or near-infrared regions of the spectrum is passed through a sample. Photons from the laser beam are absorbed by the molecules, exciting them to a virtual energy state. If the molecules relax back to the vibrational state that they started in, the reemitted photon has the same energy as the original photon. This leads to scattering of the laser light, but with no change in energy between the incoming photons and the reemitted/scattered photons. This type of scattering is known as Rayleigh scattering.

However, it is possible for the molecules to relax back to a vibrational state that is higher in energy than the state they started in. In this case, the original photon and the reemitted photon differ in energy by the amount required to vibrationally excite the molecule. Generally, the difference in energy is recorded as the difference in wavenumber ($\Delta\bar{\nu}$) between the laser light and the scattered light. A Raman spectrum is generated by plotting the intensity of the reemitted light versus . In this example the reemitted radiation is lower in energy than the incident laser light. Consequently, the change in wavenumber is positive and results in a series of peaks in the Raman spectrum known as Stokes lines.

A Raman spectrum also exhibits peaks that correspond to negative values of $\Delta\bar{\nu}$. These peaks are due to re-emitted photons that are higher in energy than the incident photons. This occurs when molecules that initially are in an excited vibrational state absorb the laser light and relax back to the lower vibrational state when they reemit the photon. These lines in the Raman spectrum are known as anti-Stokes lines. Since the Stokes lines and anti-Stokes lines gain and lose the same amount of energy, they are symmetric with respect to the peak due to elastic (Rayleigh) scattering ($\Delta\bar{\nu} = 0$). The anti-Stokes lines are appreciably less intense than the corresponding Stokes lines. This is because initially very few molecules are in excited vibrational states compared to the number in the ground state. Since anti-Stokes lines arise from the former and Stokes lines arise from the latter, the Stokes lines are much more intense. However, it should be noted that in molecules which exhibit fluorescence, the Stokes lines may be obscured while the anti-Stokes lines remain unaffected. In such cases, it is necessary to use the anti-Stokes lines despite their lower intensity.

Resonance Raman Spectroscopy

Raman spectroscopy can be used to identify chemical compounds because the values of $\Delta\bar{\nu}$ are indicative of different chemical species. This is because the frequencies of vibrational transitions depend on the atomic masses and the bond strengths. (Heavier atoms correspond to lower vibrational frequencies, while stronger bonds correspond to higher vibrational frequencies.) Thus, armed with a database of spectra from known compounds, one can unambiguously identify many different known chemical compounds based on a Raman spectrum. The number of vibrational modes scales with the number of atoms in a molecule, which means that the Raman spectra from large molecules will be very complicated. For example, proteins typically contain thousands of atoms and will therefore have thousands of vibrational modes. If these modes have similar energies ($\Delta\bar{\nu}$), then the spectrum may be incredibly cluttered and complicated.

Not all vibrational transitions will be "Raman active", i.e. some vibrational transitions will not appear in the Raman spectrum. This is because of the spectroscopic selection rules for Raman. As opposed to IR spectroscopy, where a transition can only be seen when that particular vibration causes a net change in dipole moment of the molecule, in Raman only transitions where the polarizability of the molecule changes can be observed. This is due to the fundamental difference in how IR and Raman spectroscopy access the vibrational transitions. In Raman spectroscopy, the incoming photon causes a momentary distortion of the electron distribution around a bond in a molecule, followed by reemission of the radiation as the bond returns to its normal state.

This causes temporary polarization of the bond, and an induced dipole that disappears upon relaxation. In a molecule with a centre of symmetry, a change in dipole is accomplished by loss of the centre of symmetry, while a change in polarizability is compatible with preservation of the centre of symmetry. Thus, in a centrosymmetric molecule, asymmetrical stretching and bending will be IR active and Raman inactive, while symmetrical stretching and bending will be Raman active and IR inactive. Hence, in a centrosymmetric molecule, IR and Raman spectroscopy are mutually exclusive. For molecules without a centre of symmetry, each vibrational mode may be IR active, Raman active, both, or neither. Symmetrical stretches and bends, however, tend to be Raman active.

In resonance Raman spectroscopy, the energy of the incoming laser is adjusted such that it or the scattered light coincide with an electronic transition of the molecule or crystal. In most materials the incoming and outgoing electronic resonances are sufficiently broad that they can not be distinguished. So, rather than exciting the molecule to a virtual energy state, it is excited to near one of its excited electronic transitions. Since the energy of these transitions differ from one chemical species to the next, this technique did not become applicable until the advent of tunable lasers in the early 1970s. (Tunable lasers are those where the wavelength can be altered within a specific range.) When the frequency of the laser beam is tuned to be near an electronic transition (resonance), the vibrational modes associated with that particular transition exhibit a greatly increased Raman scattering intensity. This usually overwhelms Raman signals from all of the other transitions. For instance, resonance with a π-π^* transition enhances stretching modes of the π-bonds involved with the transition, while the other modes remain unaffected.

This aspect of Raman spectroscopy becomes especially useful for large biomolecules with chromophores embedded in their structure. In such chromophores, the charge-transfer (CT) transitions of the metal complex generally enhance metal-ligand stretching modes, as well as some of modes associated with the ligands alone. Hence, in a biomolecule such as haemoglobin, tuning the laser to near the charge-transfer electronic transition of the iron centre results in a spectrum reflecting only the stretching and bending modes associated with the tetrapyrrole-iron group. Consequently, in a molecule with thousands of vibrational modes, RR spectroscopy allows us to look at relatively few vibrational modes at a time. This reduces the complexity of the spectrum and allows for easier identification of an unknown protein. Also, if a protein has more than one chromophore, different chromophores can be studied individually if their CT bands differ in energy. In addition to identifying compounds, RR spectroscopy can also supply structural identification about chromophores in some cases.

The main advantage of RR spectroscopy over traditional Raman spectroscopy is the large increase in intensity of the peaks in question (by as much as a factor of 10^6). This allows RR spectra to be generated with sample concentrations as low as 10^{-8} M. This is in stark contrast to conventional Raman spectra, which usually requires concentrations greater than 0.01 M. Also, as previously mentioned, RR spectra usually exhibit only a few peaks, and different peaks can be selected for by targeting specific electronic transitions. The main disadvantage of RR spectroscopy is the increased risk of fluorescence and photodegradation of the sample due to the increased energy of the incoming laser light. Both of these factors can be minimised by using an infrared laser instead of visible light for non resonant Raman scattering, but not in RR where the laser must be tuned to the specific resonance, unless electronic levels of lower energy are available for the system under investigation.

Instrumentation

In RR spectroscopy, the light source consists of a tunable laser, whose radiation lies in either the near-infrared, visible, or near-ultraviolet regions of the spectrum. In creating a sample handling system, RR spectroscopy offers an advantage over IR spectroscopy in that glass can be used for windows, lenses, and other optical components. Another advantage over IR spectroscopy is that whereas water absorbs strongly in the IR spectrum and may mask other signals, it only gives a weak signal in Raman spectroscopy. Therefore, water can easily be used as a solvent. Since lasers can be easily focused on small surface areas, the risk of sample heating and photodegradation is diminished, and the emitted radiation can be focused more efficiently.

Typically, the sample is placed into a tube, which can then be spun to further decrease the sample's exposure to the laser light, further diminishing the threat of photodegradation. Gaseous, liquid, and solid samples can all be analysed using RR spectroscopy. Gas and liquid samples can be put directly into the sample chamber whereas solid samples must first be ground into a powder. With gaseous and solid samples, Raman scattering may still be too weak to easily detect. For these samples, the sample holder is placed between two mirrors that reflect the laser beam multiple times through the sample.

Since scattered light leaves the sample in all directions the probes that carry the scattered light back to the detector in Raman spectroscopy may be placed at any angle. Usually, the detector probes are most placed at an angle of 135° to the path of the exiting laser light beam. Two other common arrangements position the detector probe at 90° or 180° with respect to the incident light.

Detection angles greater than 90° are generally called back-scattering detectors because they are oriented in the same direction as the incident laser light so the radiation must scatter back to the probes. In transmitting the incident laser light to the sample and the scattered light *back* to the detector, fibre-optic cables may be used. Such cables can transmit light 100 m or more, thereby allowing the analysis of samples under relatively rough experimental/environmental conditions.

After the scattered radiation exits the sample, it is sent through a monochromator. Typical monochromators consist of a diffraction grating mounted on a rotating platform. A diffraction grating causes light dispersion. Rotating the grating controls which wavelengths of scattered radiation reach the exit slit leading to the detector. The detector itself is usually a charge-coupled device (CCD), which allows the entire spectrum to be recorded simultaneously. Consequently, multiple scans can be acquired in a short period of time, which can drastically increase the signal-to-noise ratio of the spectrum.

Currently, Raman spectrometres are more expensive than more traditional dispersive instruments. As the cost of tunable lasers decrease, RR spectroscopy should see increased use, especially in the studies of metal-ligand vibrations, which reside in a region that is typically very difficult to study by other instrumental techniques. With the advent of near-infrared tunable lasers, particularly the Ti-sapphire laser (which has a range of ~700-1100 nm), Fourier Transform Resonance Raman Spectrometres may soon be commercially available. These would offer the multiplex and Jaquinot advantages of Fourier Transform (FT) techniques.

Coherent Anti-Stokes Raman Spectroscopy

Coherent anti-Stokes Raman spectroscopy, also called Coherent anti-Stokes Raman scattering spectroscopy (CARS), is a form of spectroscopy used primarily in chemistry, physics and related fields. It is sensitive to the same vibrational signatures of molecules as seen in Raman spectroscopy, typically the nuclear vibrations of chemical bonds. Unlike Raman spectroscopy, CARS employs multiple photons to address the molecular vibrations, and produces a signal in which the emitted waves are coherent with one another. As a result, CARS is orders of magnitude stronger than spontaneous Raman emission.

CARS is a third-order nonlinear optical process involving three laser beams: a pump beam of frequency ω_p, a Stokes beam of frequency ω_S and a probe beam at frequency ω_{pr}. These beams interact with the sample and generate a coherent optical signal at the anti-Stokes frequency ($\omega_{pr}+\omega_p-\omega_S$). The latter is resonantly enhanced when the frequency difference between the pump and the Stokes beams ($\omega_p-\omega_S$) coincides with the frequency of a Raman resonance, which is the basis of the technique's intrinsic vibrational contrast mechanism.

The acronym CARS, which invokes a seemingly inadvertent relation to automobiles, is actually closely related to the birth story of the technique. In 1965, a paper was published by two researchers of the Scientific Laboratory at the Ford Motor Company, P. D. Maker and R. W. Terhune, in which the CARS phenomenon was reported for the first time. Maker and Terhune used a pulsed ruby laser to investigate the third order response of several materials. They first passed the ruby beam of frequency ω through a Raman shifter to create a second beam at $\omega-\omega_v$, and then directed the two beams simultaneously onto the sample.

When the pulses from both beams overlapped in space and time, the Ford researchers observed a signal at $\omega+\omega_v$, which is the blue-shifted CARS signal. They also demonstrated that the signal increases significantly when the difference frequency ω_v between the incident beams matches a Raman frequency of sample. Maker and Terhune called their technique simply 'three wave mixing experiments'. The name coherent anti-Stokes Raman spectroscopy was assigned almost ten years later, by Begley et al. at Stanford University in 1974. Since then, this vibrationally sensitive nonlinear optical technique has been commonly known as CARS.

Principle

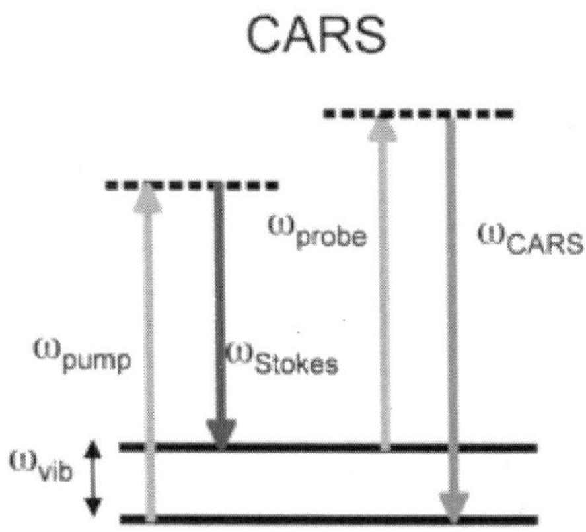

Figure: CARS energy diagram

The CARS process can be physically explained by using either a classical oscillator model or by using a quantum mechanical model that incorporates the energy levels of the molecule. Classically, the Raman active vibrator is modelled as a (damped) harmonic oscillator with a characteristic frequency of ω_v. In CARS, this oscillator is not driven by a single optical wave, but by the difference frequency (ω_p-ω_S) between the pump and the Stokes beams instead. This driving mechanism is similar to hearing the low combination tone when striking two different high tone piano keys: your ear is sensitive to the difference frequency of the high tones. Similarly, the Raman oscillator is susceptible to the difference frequency of two optical waves. When the difference frequency ω_p-ω_S approaches ω_v, the oscillator is driven very efficiently. On a molecular level, this implies that the electron cloud surrounding the chemical bond is vigorously oscillating with the frequency ω_p-ω_S. These electron motions alter the optical properties of the sample, i.e. there is a periodic modulation of the refractive index of the material. This periodic modulation can be probed by a third laser beam, the probe beam. When the probe beam is propagating through the periodically altered medium, it acquires the same modulation. Part of the probe, originally at ω_{pr} will now get modified to ω_{pr}+ω_p-ω_S, which is the observed anti-Stokes emission. Under certain beam geometries, the anti-Stokes emission may diffract away from the probe beam, and can be detected in a separate direction.

While intuitive, this classical picture does not take into account the quantum mechanical energy levels of the molecule. Quantum mechanically, the CARS process can be understood as follows. Our molecule is initially in the ground state, the lowest energy state of the molecule. The pump beam excites the molecule to a virtual state. A virtual state is not an eigenstate of the molecule and it can not be occupied but it does allow for transitions between otherwise uncoupled real states. If a Stokes beam is simultaneously present along with the pump, the virtual state can be used as an instantaneous gateway to address a vibrational eigenstate of the molecule. The joint action of the pump and the Stokes has effectively established a coupling between the ground state and the vibrationally excited state of the molecule.

The molecule is now in two states at the same time: it resides in a coherent superposition of states. This coherence between the states can be probed by the probe beam, which promotes the system to a virtual state. Again, the molecule cannot stay in the virtual state and will fall back instantaneously to the ground state under the emission of a photon at the anti-Stokes frequency. The molecule is no longer in a superposition, as it resides again in one state, the ground state. In the quantum mechanical model, no energy is deposited in the molecule during the CARS process. Instead, the molecule acts like a medium for converting the frequencies of the three incoming waves into a CARS signal (a parametric process). There are, however, related coherent Raman process that occur simultaneously which do deposit energy into the molecule.

Comparison to Raman Spectroscopy

CARS is often compared to Raman spectroscopy as both techniques probe the same Raman active modes. Raman can be done using a single continuous wave (CW) laser whereas CARS requires (generally) two pulsed laser sources. The Raman signal is detected on the red side of the incoming radiation where it might have to compete with other fluorescent processes. The CARS signal is detected on the blue side, which is free from fluorescence, but it comes with a non-resonant contribution. The differences between the signals from Raman and CARS (there are many variants of both techniques) stems largely from the fact that Raman relies on a spontaneous transition whereas CARS relies on a coherently driven transition. The total Raman signal collected from a sample is the incoherent addition of the signal from individual molecules. It is therefore linear in the concentration of those molecules and the signal is emitted in all directions. The total

CARS signal comes from a coherent addition of the signal from individual molecules. For the coherent addition to be additive, phase-matching must be fulfilled. For tight focusing conditions this is generally not a restriction. Once phase-matching is fulfilled the signal amplitude grows linear with distance so that the power grows quadratically. This signal forms a collimated beam that is therefore easily collected. The fact that the CARS signal is quadratic in the distance makes it quadratic with respect to the concentration and therefore especially sensitive to the majority constituent. The total CARS signal also contains an inherent non-resonant background. This non-resonant signal can be considered as the result of (several) far off-resonance transitions that also add coherently. The resonant amplitude contains a phase shift of Pi over the resonance whereas the non-resonant part does not. The spectral line shape of the CARS intensity therefore resembles a Fano-profile which is shifted with respect to the Raman signal. To compare the spectra from multi-component compounds, the (resonant) CARS spectral amplitude should be compared to the Raman spectral intensity.

Theoretically Raman spectroscopy and CARS spectroscopy are equally sensitive as they use the same molecular transitions. However, given the limits on input power (damage threshold) and detector noise (integration time), the signal from a single transition can be collected much faster in practical situation (a factor of 10^5) using CARS. Imaging of known substances (known spectra) is therefore often done using CARS. Given the fact that CARS is a higher order nonlinear process, the CARS signal from a single molecule is larger than the Raman signal from a single molecule for a sufficiently high driving intensity. However at very low concentrations, the advantages of the coherent addition for CARS signal reduces and the presence of the incoherent background becomes an increasing problem.

Since CARS is such a nonlinear process there are not really any 'typical' experimental numbers. One example is given below under the explicit warning that just changing the pulse duration by one order of magnitude changes the CARS signal by three orders of magnitude. The comparison should only be used as an indication of the order of magnitude of the signals. 200 mW average power input (CW for the Raman), in a 0.9NA objective with a centre wavelength around 800 nm, constitutes a power density of 26 MW/cm,^{2} (focus length = 1.5 micrometre, focus volume = 1.16 micrometre^3, photon energy = 2.31

10^{-19} J or 1.44 eV). The Raman cross section for the vibration of the aromatic ring in Toluene around 1000 cm^{-1} is on the order of $10^{-29}cm^2$/molecule*steradian. Therefore the Raman signal is around 26 10^{-22} W/molecule*steradian or 3.3 10^{-21} W/molecule (over 4Pi). That is 0.014 photon/sec*molecule. The density of Toluene = 0.8668 10^3 kg/m^3, Molecular mass = 92.14 10^{-3} kg/mol. Therefore the focal volume (~1 cubic micrometre) contains 6 10^9 molecules. Those molecules together generate a Raman signal in the order of 2 10^{-11} W (20pW) or roughly one hundred million photons/sec (over a 4Pi solid angle). A CARS experiment with similar parameters (150 mW at 1064 nm, 200 mW at 803.5 nm, 15ps pulses at 80Mhz repetition frequency, same objective lens) yields roughly 17.5 10^{-6} W (on the 3000 cm^{-1} line, which has 1/3 of the strength and roughly 3 times the width). This CARS power is roughly 10^6 higher than the Raman but since there are 6 10^9 molecules, the signal per molecule from CARS is only 4 10^{-25} W/molecule*sec or 1.7 10^{-6} photons/molecule*sec. If we allow two factors of three (line strength and line width) then the spontaneous Raman signal per molecule still exceeds the CARS per molecule by more than two orders of magnitude. The coherent addition of the CARS signal from the molecules however yields a total signal that is much higher than the Raman.

The sensitivity in many CARS experiments is not limited by the detection of CARS photons but rather by the distinction between the resonant and non-resonant part of the CARS signal.

Applications

CARS is used for species selective microscopy and combustion diagnostics. The first exploits the selectivity of vibrational spectroscopy whereas the latter is aimed at temperature measurements; the CARS signal is temperature dependent. The strength of the signal scales (non-linearly) with the difference in the ground state population and the vibrationally excited state population. Since the population of states follows the temperature dependent Boltzmann Distribution, the CARS signal carries an intrinsic temperature dependence as well. This temperature dependence makes CARS a popular technique for monitoring the temperature of hot gases and flames. More recently, CARS has been utilised as a method for non-invasive imaging of lipids in biological samples, both *in vivo* and *in vitro*. CARS-based detectors for roadside bombs are under development.

Chapter 6

Infrared Spectroscopy

Infrared spectroscopy (IR spectroscopy) is the spectroscopy that deals with the infrared region of the electromagnetic spectrum, that is light with a longer wavelength and lower frequency than visible light. It covers a range of techniques, mostly based on absorption spectroscopy. As with all spectroscopic techniques, it can be used to identify and study chemicals. A common laboratory instrument that uses this technique is a Fourier transform infrared (FTIR) spectrometre.

The infrared portion of the electromagnetic spectrum is usually divided into three regions; the near-, mid- and far- infrared, named for their relation to the visible spectrum. The higher-energy near-IR, approximately 14000–4000 cm^{-1} (0.8–2.5 μm wavelength) can excite overtone or harmonic vibrations. The mid-infrared, approximately 4000–400 cm^{-1} (2.5–25 μm) may be used to study the fundamental vibrations and associated rotational-vibrational structure. The far-infrared, approximately 400–10 cm^{-1} (25–1000 μm), lying adjacent to the microwave region, has low energy and may be used for rotational spectroscopy. The names and classifications of these subregions are conventions, and are only loosely based on the relative molecular or electromagnetic properties.

Infrared spectroscopy exploits the fact that molecules absorb specific frequencies that are characteristic of their structure. These absorptions are resonant frequencies, i.e. the frequency of the absorbed radiation matches the frequency of the bond or group that vibrates. The energies are determined by the shape of the molecular potential energy surfaces, the masses of the atoms, and the associated vibronic coupling.

In particular, in the Born–Oppenheimer and harmonic approximations, i.e. when the molecular Hamiltonian corresponding to the electronic ground state can be approximated by a harmonic oscillator in the neighbourhood of the equilibrium molecular geometry, the resonant frequencies are determined by the normal modes corresponding to the molecular electronic ground state potential energy surface. Nevertheless, the resonant frequencies can be in a first approach related to the strength of the bond, and the mass of the atoms at either end of it. Thus, the frequency of the vibrations can be associated with a particular bond type.

Number of Vibrational Modes

In order for a vibrational mode in a molecule to be "IR active,– it must be associated with changes in the dipole. A permanent dipole is not necessary, as the rule requires only a change in dipole moment.

A molecule can vibrate in many ways, and each way is called a *vibrational mode.* For molecules with N atoms in them, linear molecules have 3N – 5 degrees of vibrational modes, whereas nonlinear molecules have 3N – 6 degrees of vibrational modes (also called vibrational degrees of freedom). As an example H_2O, a non-linear molecule, will have $3 \times 3 - 6 = 3$ degrees of vibrational freedom, or modes.

Simple diatomic molecules have only one bond and only one vibrational band. If the molecule is symmetrical, e.g. N_2, the band is not observed in the IR spectrum, but only in the Raman spectrum. Asymmetrical diatomic molecules, e.g. CO, absorb in the IR spectrum. More complex molecules have many bonds, and their vibrational spectra are correspondingly more complex, i.e. big molecules have many peaks in their IR spectra.

Special Effects

The simplest and most important IR bands arise from the "normal modes," the simplest distortions of the molecule. In some cases, "overtone bands" are observed. These bands arise from the absorption of a photon that leads to a doubly excited vibrational state. Such bands appear at approximately twice the energy of the normal mode. Some vibrations, so-called 'combination modes," involve more than one normal mode. The phenomenon of Fermi resonance can arise when two modes are similar in energy; Fermi resonance results in an unexpected shift in energy and intensity of the bands.

Practical IR Spectroscopy

The infrared spectrum of a sample is recorded by passing a beam of infrared light through the sample. When the frequency of the IR is the same as the vibrational frequency of a bond, absorption occurs. Examination of the transmitted light reveals how much energy was absorbed at each frequency (or wavelength). This can be achieved by scanning the wavelength range using a monochromator. Alternatively, the whole wavelength range is measured at once using a Fourier transform instrument and then a transmittance or absorbance spectrum is generated using a dedicated procedure. Analysis of the position, shape and intensity of peaks in this spectrum reveals details about the molecular structure of the sample.

This technique works almost exclusively on samples with covalent bonds. Simple spectra are obtained from samples with few IR active bonds and high levels of purity. More complex molecular structures lead to more absorption bands and more complex spectra. The technique has been used for the characterisation of very complex mixtures. Spectra issues with Infrared Flourescense are rare.

Sample Preparation

Gaseous samples require a sample cell with a long pathlength to compensate for the diluteness. The pathlength of the sample cell depends on the concentration of the compound of the interest. A simple glass tube with length of 5 to 10 cm equipped with infrared windows at the both ends of the tube can be used for concentrations down to several hundred ppm. Sample gas concentrations well below ppm can be measured with a White's cell in which the infrared light is guided with mirrors to travel through the gas. White's cells are available with optical pathlength starting from 0.5 m up to hundred metres.

Liquid samples can be sandwiched between two plates of a salt (commonly sodium chloride, or common salt, although a number of other salts such as potassium bromide or calcium fluoride are also used). The plates are transparent to the infrared light and do not introduce any lines onto the spectra.

Solid samples can be prepared in a variety of ways. One common method is to crush the sample with an oily mulling agent (usually Nujol) in a marble or agate mortar, with a pestle. A thin film of the mull is smeared onto salt plates and measured. The second method is to grind a quantity of the sample with a specially purified salt (usually potassium bromide) finely (to remove scattering effects from large crystals). This powder mixture is then pressed in a mechanical press to form a translucent pellet through which the beam of the spectrometre can pass.

A third technique is the "cast film" technique, which is used mainly for polymeric materials. The sample is first dissolved in a suitable, non hygroscopic solvent. A drop of this solution is deposited on surface of KBr or NaCl cell. The solution is then evaporated to dryness and the film formed on the cell is analysed directly. Care is important to ensure that the film is not too thick otherwise light cannot pass through. This technique is suitable for qualitative analysis. The final method is to use microtomy to cut a thin (20–100 μm) film from a solid sample. This is one of the most important ways of analysing failed plastic products for example because the integrity of the solid is preserved.

In photoacoustic spectroscopy the need for sample treatment is minimal. The sample, liquid or solid, is placed into the sample cup which is inserted into the photoacoustic cell which is then sealed for the measurement. The sample may be one solid piece, powder or basically in any form for the measurement. For example, a piece of rock can be inserted into the sample cup and the spectrum measured from it.

It is important to note that spectra obtained from different sample preparation methods will look slightly different from each other due to differences in the samples' physical states.

Comparing to a Reference

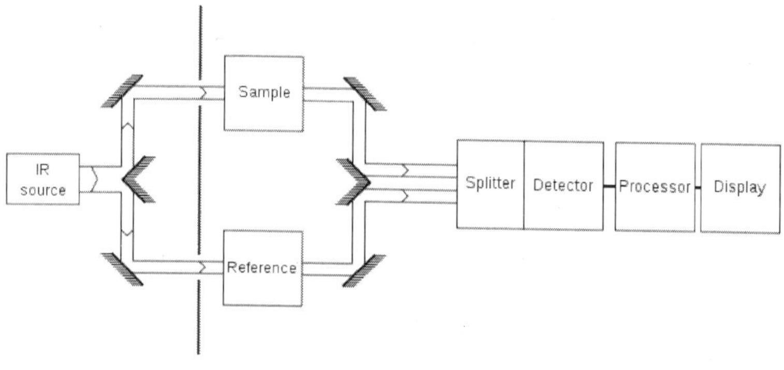

Figure:

Schematics of a two-beam absorption spectrometre. A beam of infrared light is produced, passed through an interferometre (not shown), and then split into two separate beams. One is passed through the sample, the other passed through a reference. The beams are both reflected back towards a detector, however first they pass through a splitter, which quickly alternates which of the two beams enters the detector. The two signals are then compared and a printout is obtained. This "two-beam" setup gives accurate spectra even if the intensity of the light source drifts over time.

To take the infrared spectrum of a sample, it is necessary to measure both the sample and a "reference" (or "control"). This is because each measurement is affected by not only the light-absorption properties of the sample, but also the properties of the instrument (for example, what light source is used, what infrared detector is used, etc.). The reference measurement makes it possible to eliminate the instrument influence. Mathematically, the sample transmission spectrum is divided by the reference transmission spectrum.

The appropriate "reference" depends on the measurement and its goal. The simplest reference measurement is to simply remove the sample (replacing it by air). However, sometimes a different reference is more useful. For example, if the sample is a dilute solute dissolved in water in a beaker, then a good reference measurement might be to measure pure water in the same beaker. Then the reference measurement would cancel out not only all the instrumental properties (like what light source is used), but also the light-absorbing and light-reflecting properties of the water and beaker, and the final result would just show the properties of the solute (at least approximately).

A common way to compare to a reference is sequentially: first measure the reference, then replace the reference by the sample and measure the sample. This technique is not perfectly reliable; if the infrared lamp is a bit brighter during the reference measurement, then a bit dimmer during the sample measurement, the measurement will be distorted. More elaborate methods, such as a "two-beam" setup, can correct for these types of effects to give very accurate results. The Standard addition method can be used to statistically cancel these errors.

FTIR

An interferogram from an FTIR measurement. The horizontal axis is the position of the mirror, and the vertical axis is the amount of light detected. This is the "raw data" which can be Fourier transformed to get the actual spectrum.

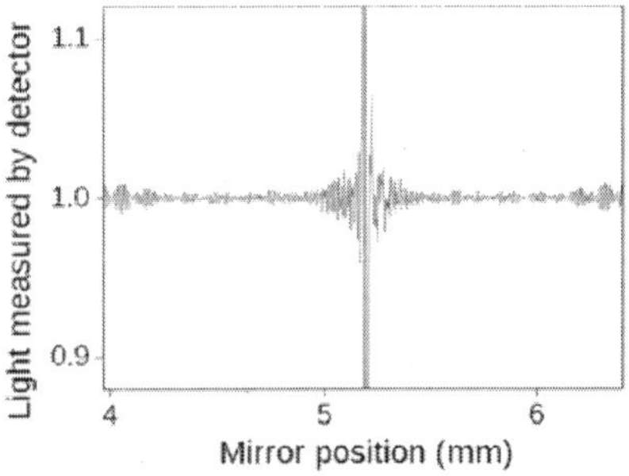

Figure:

Fourier transform infrared (FTIR) spectroscopy is a measurement technique that allows one to record infrared spectra. Infrared light is guided through an interferometre and then through the sample (or vice versa). A moving mirror inside the apparatus alters the distribution of infrared light that passes through the interferometre. The signal directly recorded, called an "interferogram", represents light output as a function of mirror position. A data-processing technique called Fourier transform turns this raw data into the desired result (the sample's spectrum): Light output as a function of infrared wavelength (or equivalently, wavenumber). As described above, the sample's spectrum is always compared to a reference.

There is an alternate method for taking spectra (the "dispersive" or "scanning monochromator" method), where one wavelength at a time passes through the sample. The dispersive method is more common in UV-Vis spectroscopy, but is less practical in the infrared than the FTIR method. One reason that FTIR is favoured is called "Fellgett's advantage" or the "multiplex advantage": The information at all frequencies is collected simultaneously, improving both speed and signal-to-noise ratio. Another is called "Jacquinot's Throughput Advantage": A dispersive measurement requires detecting much lower light levels than an FTIR measurement. There are other advantages, as well as some disadvantages, but virtually all modern infrared spectrometres are FTIR instruments.

Absorption Bands

Infrared spectroscopy is a simple and reliable technique widely used in both organic and inorganic chemistry, in research and industry. It is used in quality control, dynamic measurement, and monitoring applications such as the long-term unattended measurement of CO_2

concentrations in greenhouses and growth chambers by infrared gas analysers.

It is also used in forensic analysis in both criminal and civil cases, for example in identifying polymer degradation. It can be used in detecting how much alcohol is in the blood of a suspected drunk driver measured as 1/10,000 g/mL = 100 μg/mL.

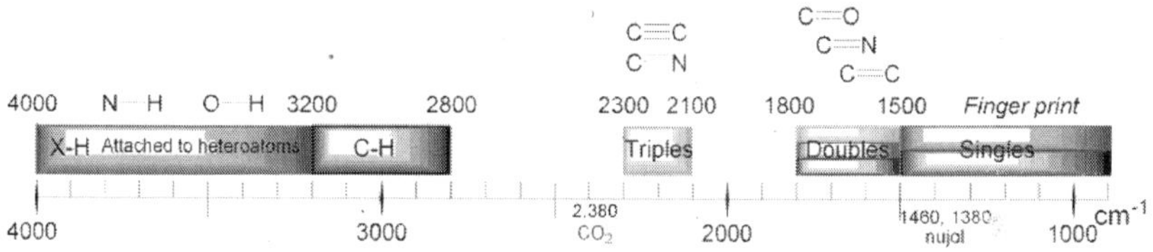

Figure: Wavenumbers listed in cm^{-1}.

A useful way of analysing solid samples without the need for cutting samples uses ATR or attenuated total reflectance spectroscopy. Using this approach, samples are pressed against the face of a single crystal. The infrared radiation passes through the crystal and only interacts with the sample at the interface between the two materials.

With increasing technology in computer filtering and manipulation of the results, samples in solution can now be measured accurately (water produces a broad absorbance across the range of interest, and thus renders the spectra unreadable without this computer treatment).

Some instruments will also automatically tell you what substance is being measured from a store of thousands of reference spectra held in storage. Infrared spectroscopy is also useful in measuring the degree of polymerization in polymer manufacture. Changes in the character or quantity of a particular bond are assessed by measuring at a specific frequency over time. Modern research instruments can take infrared measurements across the range of interest as frequently as 32 times a second. This can be done whilst simultaneous measurements are made using other techniques. This makes the observations of chemical reactions and processes quicker and more accurate.

Infrared spectroscopy has also been successfully utilised in the field of semiconductor microelectronics: for example, infrared spectroscopy can be applied to semiconductors like silicon, gallium arsenide, gallium nitride, zinc selenide, amorphous silicon, silicon nitride, etc. The instruments are now small, and can be transported, even for use in field trials.

Isotope Effects

The different isotopes in a particular species may give fine detail in infrared spectroscopy. For example, the O–O stretching frequency (in reciprocal centimetres) of oxyhemocyanin is experimentally determined to be 832 and 788 cm^{-1} for ν(^{16}O–^{16}O) and ν(^{18}O–^{18}O), respectively.

By considering the O–O bond as a spring, the wavenumber of absorbance, ν can be calculated:

$$\nu = \frac{1}{2\pi c}\sqrt{\frac{k}{\mu}}$$

where k is the spring constant for the bond, c is the speed of light, and μ is the reduced mass of the A–B system:

$$\mu = \frac{m_A m_B}{m_A + m_B}$$

(m_i is the mass of atom i).

The reduced masses for ^{16}O–^{16}O and ^{18}O–^{18}O can be approximated as 8 and 9 respectively. Thus

$$\frac{\nu(^{16}O)}{\nu(^{18}O)} = \sqrt{\frac{9}{8}} \approx \frac{832}{788}.$$

Where ν is the wavenumber; [wavenumber = frequency/(speed of light)]

The effect of isotopes, both on the vibration and the decay dynamics, has been found to be stronger than previously thought. In some systems, such as silicon and germanium, the decay of the anti-symmetric stretch mode of interstitial oxygen involves the symmetric stretch mode with a strong isotope dependence. For example, it was shown that for a natural silicon sample, the lifetime of the anti-symmetric vibration is 11.4 ps. When the isotope of one of the silicon atoms is increased to ^{29}Si, the lifetime increases to 19 ps. In similar manner, when the silicon atom is changed to ^{30}Si, the lifetime becomes 27 ps.

Two-Dimensional IR

Two-dimensional infrared correlation spectroscopy analysis is the application of 2D correlation analysis on infrared spectra. By extending the spectral information of a perturbed sample, spectral analysis is simplified and resolution is enhanced. The 2D synchronous and 2D asynchronous spectra represent a graphical overview of the spectral changes due to a perturbation (such as a changing concentration or changing temperature) as well as the relationship between the spectral changes at two different wavenumbers.

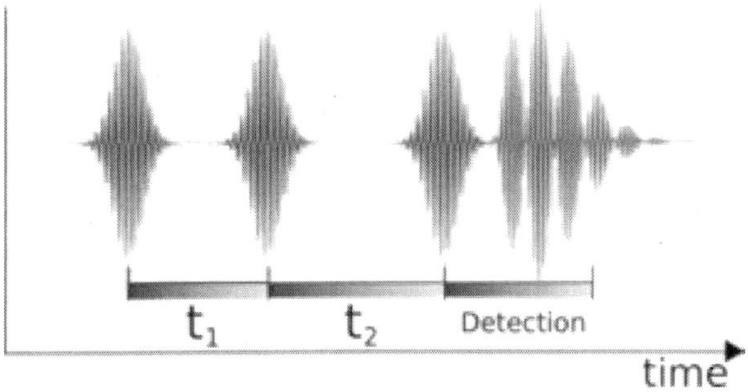

Figure:

Pulse Sequence used to obtain a two-dimensional Fourier transform infrared spectrum. The time period $\tau_1$is usually referred to as the coherence time and the second time period $\tau_2$is known as the waiting time. The excitation frequency is obtained by Fourier transforming along the axis.

Nonlinear two-dimensional infrared spectroscopy is the infrared version of correlation spectroscopy. Nonlinear two-dimensional infrared spectroscopy is a technique that has become available with the development of femtosecond infrared laser pulses. In this experiment, first a set of pump pulses are applied to the sample. This is followed by a waiting time, wherein the system is allowed to relax. The typical waiting time lasts from zero to several picoseconds, and the duration can be controlled with a resolution of tens of femtoseconds. A probe pulse is then applied resulting in the emission of a signal from the sample. The nonlinear two-dimensional infrared spectrum is a two-dimensional correlation plot of the frequency ω_1 that was excited by the initial pump pulses and the frequency ω_3 excited by the probe pulse after the waiting time. This allows the observation of coupling between different vibrational modes; because of its extremely high time resolution, it can be used to monitor molecular dynamics on a picosecond timescale. It is still a largely unexplored technique and is becoming increasingly popular for fundamental research.

As with two-dimensional nuclear magnetic resonance (2DNMR) spectroscopy, this technique spreads the spectrum in two dimensions and allows for the observation of cross peaks that contain information on the coupling between different modes. In contrast to 2DNMR, nonlinear two-dimensional infrared spectroscopy also involves the excitation to overtones. These excitations result in excited state absorption peaks located below the diagonal and cross peaks. In 2DNMR, two distinct techniques, COSY and NOESY, are frequently used. The cross peaks in the first are related to the scalar coupling, while in the latter they are related to the spin transfer between different nuclei. In nonlinear two-dimensional infrared spectroscopy, analogs have been drawn to these 2DNMR techniques. Nonlinear two-dimensional infrared spectroscopy with zero waiting time corresponds to COSY, and nonlinear two-dimensional infrared spectroscopy with finite waiting time allowing vibrational population transfer corresponds to NOESY. The COSY variant of nonlinear two-dimensional infrared spectroscopy has been used for determination of the secondary structure content proteins.

Chapter 7

Electronic Spectroscopy

Often, during electronic spectroscopy, the electron is excited first from an initial low energy state to a higher state by absorbing photon energy from the spectrophotometre. If the wavelength of the incident beam has enough energy to promote an electron to a higher level, then we can detect this in the absorbance spectrum.

Once in the excited state, the electron has higher potential energy and will relax back to a lower state by emitting photon energy. This is called fluorescence and can be detected in the spectrum as well.

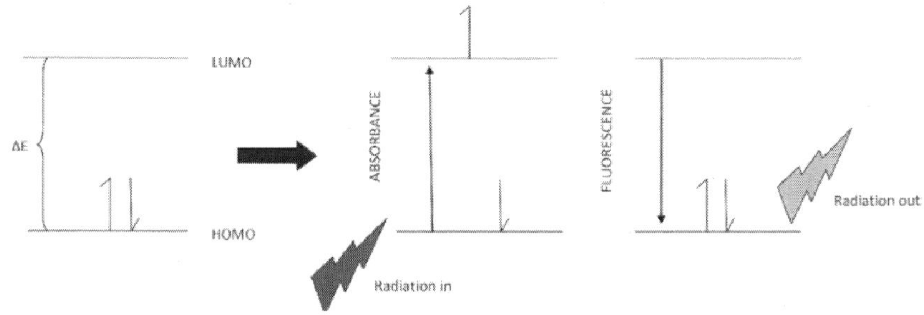

Embedded into the electronic states (n=1,2,3...) are vibrational levels (v=1,2,3...) and within these are rotational energy levels (j=1,2,3...). Often, during electronic transitions, the initial state may have the electron in a level that is excited for both vibration and rotation. In other words, n=0, v does not = 0 and r does not =0. This can be true for the ground state and the excited state. In addition, due to the Frank Condon Factor, which describes the overlap between vibrational states of two electronic states, there may be visible vibrational bands within the absorption bands. Therefore, vibrational fine structure that can be seen in the absorption spectrum gives some indication of the degree of Frank Condon overlap between electronic states.

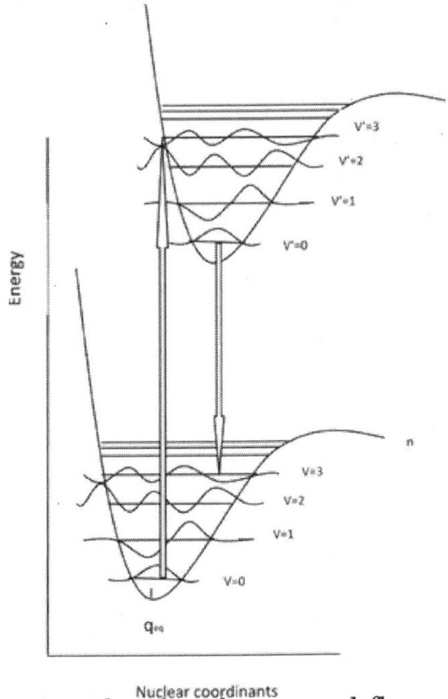

When interpreting the absorbance and fluorescence spectra of a given molecule, compound, material, or an elemental material, understanding the possible electronic transitions is crucial. Assigning the peaks in the absorption spectrum can become easier when considering which transitions are allowed by symmetry, the Laporte Rules, electron spin, or vibronic coupling. Knowing the degree of allowedness, one can estimate the intensity of the transition, and the extinction coefficient associated with that transition. These guidelines are a few examples of the selection rules employed for interpreting the origin of spectral bands.

Only a complete model of molecular energy diagrams for the species under investigation can make clear the possible electronic transitions. Every different compound will have unique energy spacing between electronic levels, and depending on the type of compound, one can categorize these spacings and find some commonality. For example, aromatic compounds pi to pi* and n to pi* transitions where as inorganic compounds can have similar transitions with Metal to Ligand Charge Transfer (MLCT) and Ligand to Metal Charge Transfer (LMCT) in addition to d-d transitions, which lead to the bright colours of transition metal complexes. Although surprises in science often lead to discovery, it is more fortuitous for the interpreter to predict the spectra rather than being baffled by the observation. The following section will discuss the interpretation of electronic absorption spectra given the nature of the chemical species being studied. This includes an understanding of the molecular or elemental electronic state symmetries, Russell-Sanders states, spin multiplicities, and forbidden and allowed transitions of a given species.

As the light passes through the monochrometre of the spectrophotometre, it hits the sample with some wavelength and corresponding energy. The ratio of the initial intensity of this light and the final intensity after passing through the sample is measured and recorded as absorbance (Abs). When absorbance is measured at different wavelengths, an absorbance spectrum of Abs vs wavelength can be obtained. This spectra reveals the wavelengths of light that are absorbed by the chemical specie, and is specific for each different chemical. Many electronic transitions can be visible in the spectrum if the energy of the incident light matches or surpasses the quantum of energy separating the ground state and that particular excited state. An example of an absorbance spectrum is given below.

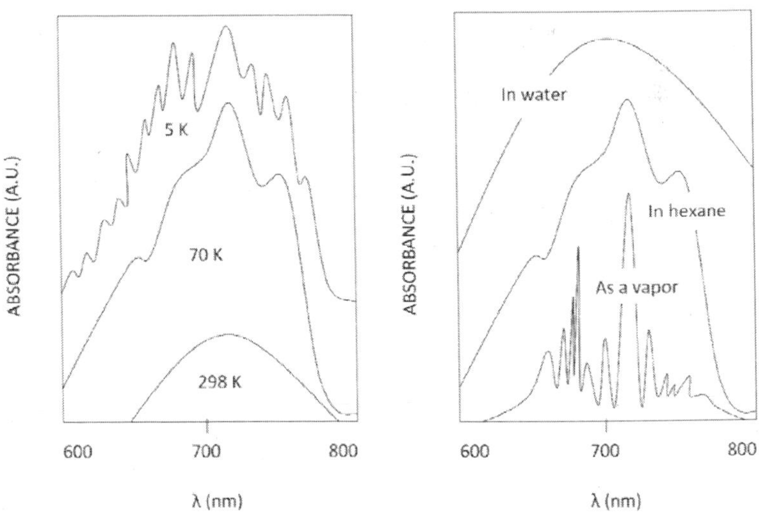

Figure:

Temperature Effects

Here we can see the effect of temperature and also the effect of solvents on the clarity of the spectrum. We can see from these anonymous compounds that decreasing the temperature allows the vibrational fine structure to emerge. These vibrational bands embedded within the electronic bands represent the transitions from v=n to v'=n. Generally, the v=0 to v'=0 transition is the one with the lowest frequency. From there, increasing energy, the transitions can be from v=0 to v'=n, where n=1,2,3... With a higher temperature, the vibrational transitions become averaged in the spectrum due to the presence of vibrational hot bands and Fermi Resonance, and with this, the vibrational fine structure is lost at higher temperatures.

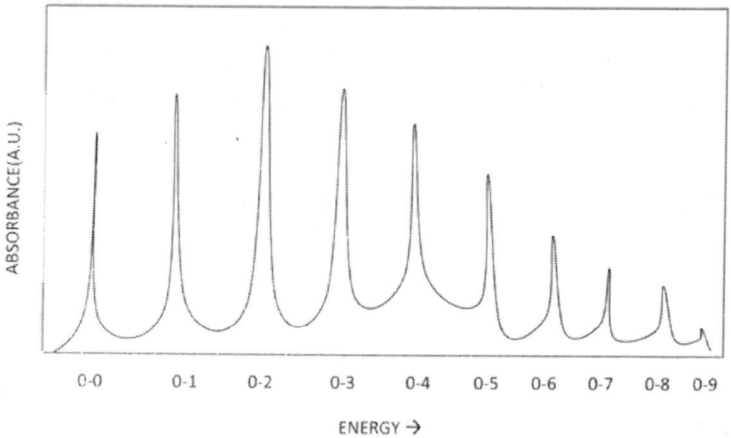

Figure:

On the left, we see the electronic absorption of an anonymous diatomic molecule. The bands represent the jumping of the electron from the ground electronic state/ground vibrational state n=0/v=0 to the excited electronic state/ground vibrational state n=1/v=0. This is denoted as 0-0 and is defined as the lowest energy transition. The other transitions involve an electronic transition in combination with vibrational transitions within the excited electronic state. In other words, the bands labelled 0-1, 0-2,...etc represent the electrons that travel higher in energy than the purely electronic transition (0-0).

This introduces the possibility of vibrational information within the electronic spectrum. In fact, due to the Frank Condon overlap of the ground state and excited state, the 0-2 transition is the most probable of all of the transitions. We can see this in the spectra as the 0-2 has the greatest absorbance. If the potential energy wells of the ground and excited states were perfectly eclipsed on top of one another, then the 0-0 transition would be the most probable transition with the highest intensity band. This however, as we have seen, is not often the case. In deed the equilibrium bond length tends to increase as the energy of the state increases usually due to antibonding nature of the state increasing the intramolecular bond length. This causes increased Frank Condon overlap for some transitions and decreased for others.

Solvent Effects

The effect that the solvent plays on the absorption spectrum is also very important. It is clear that polar solvents give rise to broad bands, non-polar solvents show more resolution, though, completely removing the solvent gives the best resolution. This is due to solvent-solute interaction. The solvent can interact with the solute in its ground state or excited state through intermolecular bonding. For example, a polar solvent like water has the ability of hydrogen bonding with the solute if the solute has a hydrogen bonding component, or simply through induced dipole-dipole interactions. The non-polar solvents can interact though polarizability via London interactions also causing a blurring of the vibronic manifold. This is due to the solvent's tendency to align its dipole moment with the dipole moment of the solute. Depending on the interaction, this can cause the ground state and the excited state of the solute to increase or decrease, thus changing the frequency of the absorbed photon. Due to this, there are many different transition energies that become average together in the spectra. This causes peak-broadening. The effects of peak broadening are most severe for polar solvent, less so for non-polar solvents, and absent when the solute is in vapour phase.

Group Theory and The Transition Moment Integral

When estimating the intensities of the absorption peaks, we use the molar absorptivity constant (epsilon). If the transition is "allowed" then the molar absorptivity constant from the Beer's Law Plot will be high. This means that the probability of transition is large. If the transition is not allowed, then there will be no intensity and no peak on the spectrum. Transitions can be "partially allowed" as well, and these bands appear with a lower intensity than the full allowed transitions. One way to decide whether a transition will be allowed or not is to use symmetry arguments with Group Theory.

If the symmetries of the ground and final state of a transition are correct, then the transition is symmetry allowed. We express this by modifying the transition moment integral from an integral of eigenstates to an orthogonally expressed direct product of the symmetries of the states. The conversions of integration to direct products of symmetry as shown gives spectroscopists a short cut into deciding whether the transition will be allowed or forbidden. A transition will be forbidden if the direct products of the symmetries of the electronic states with the coupling operator are odd. More specifically, if the direct product does not contain the totally symmetric representation, then the transition is forbidden by symmetry arguments. If the product does contain the totally symmetric representation (A, A_1, A_{1g}...etc) then the transition is symmetry allowed.

Some transitions are forbidden by the equation (1) and one would not expect to be able to see the band that corresponds to the transition; however, a weak absorbance band is quite clear on the spectrum of many compounds. The transition may be forbidden via pure electronic symmetries; however, for an octahedral complex for example since it has a centre of inversion, the transition is weakly allowed because of vibronic coupling. When the octahedra of a transition metal complex is completely symmetric (without vibrations), the transition cannot occur. However, when vibrations exist, they temporarily perturb the symmetry of the complex and allow the transition by equation (2). If the product of all of these representations contains the totally symmetric representation, then the transition will be allowed via vibronic coupling even if it forbidden electronically.

Hot Bands

Some transitions are forbidden by symmetry and do not appear in the absorption spectrum. If the symmetries are correct, then another state besides the ground state can be used to make the otherwise forbidden transition possible. This is accomplished by hot bands, meaning the electrons in the ground state are heated to a higher energy level that has a different symmetry. When the transition moment integral is solved with the new hot ground state, then the direct product of the symmetries may contain the totally symmetric representation. If we employ the old saying, "You can't get there from here!" then we would be referring to the transition from the ground state to the excited state. However, if we thermally excite the molecules from out of the ground state, then, "we can get there from here!"

Knowing whether a transition will be allowed by symmetry is an essential component to interpreting the spectrum. If the transition is allowed, then it should be visible with a large extinction coefficient. If it is forbidden, then it should only appear as a weak band if it is allowed by vibronic coupling. In addition to this, a transition can also be spin forbidden. The examples below of excited state symmetries, give an indication of what spin forbidden means:

$^1A_{1g}$ ⟶ $^3E_{1u}$ (spin forbidden)

$^1A_{1g}$ ⟶ $^3B_{1u}$ (spin forbidden)

$^1A_{1g}$ ⟶ $^3B_{2u}$ (spin forbidden)

$^1A_{1g}$ ⟶ $^1E_{1u}$ (spin allowed)

$^1A_{1g}$ ⟶ $^1B_{1u}$ (spin allowed)

$^1A_{1g}$ ⟶ $^1B_{2u}$ (spin allowed)

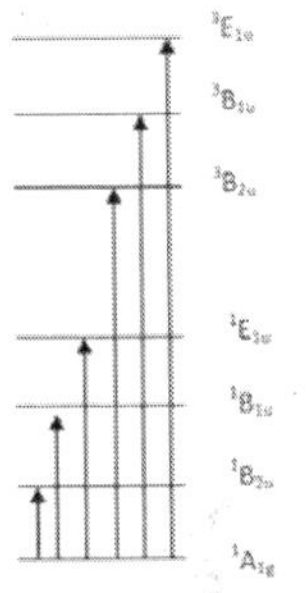

States from first excited state....$(e_{1g})^3(e_{2u})^1$
Direct product = $B_{1u} + B_{2u} + E_{1u}$
Spin multiplicity from S=1 and S=0

These states are derived from the electron configuration of benzene. Once we have the molecular orbital energy diagram for benzene, we can assign symmetries to each orbital arrangement of the ground state. From here, we can excite an electron from the Highest Occupied Molecular Orbital (HOMO) to the Lowest Unoccupied Molecular Orbital (LUMO). This is the lowest energy transition. Other transitions include moving the electron above the LUMO to higher energy molecular orbitals. To solve for the identity of the symmetry of the excited state, one can take the direct product of the HOMO symmetry and the excited MO symmetry. This give a letter (A, B, E..) an the subscript (1u, 2u, 1g...). The superscript is the spin multiplicity, and from single electron transitions, the spin multiplicity is 2S+1 = M, where S = 1 with two unpaired electrons having the same spin and S=0 when the excited electron flips its spin so that the two electrons have opposite spin. This gives M=1 and M=3 for benzene above. From the results above, we have three transitions that are spin allowed and three that are spin forbidden.

Once we take the direct product of the symmetries and the coupling operator for each of these states given above, we find that only the A_{1g} to E_{1u} transition is allowed by symmetry. Therefore, we have information regarding spin and symmetry allowedness and we have an idea of what the spectra will look like:

When interpreting the spectrum, it is clear that some transitions are more probable than others. According to the symmetry of excited states, we can now order them from low energy to high energy based on the position of the peaks (E1u is the highest, then B1u, and B2u is lowest). The A_{1g} to E_{1u} transition is fully allowed and therefore the most intense peak. The A_{1g} to B_{1u} and A_{1g} to B_{2u} transitions are symmetry forbidden and thus have a lower probability which is evident from the lowered intensity of their bands. The singlet A_{1g} to triplet B_{1u} transition is both symmetry forbidden and spins forbidden and therefore has the lowest intensity. This transition is forbidden by spin arguments; however, a phenomenon known as spin-orbit coupling can allow this transition to be weakly allowed as well. If spin-orbit coupling exists, then the singlet state has the same total angular momentum as the triplet state so the two states can interact. A small amount of singlet character in the triplet state leads to a transition moment integral that is non-zero, so the transition is allowed.

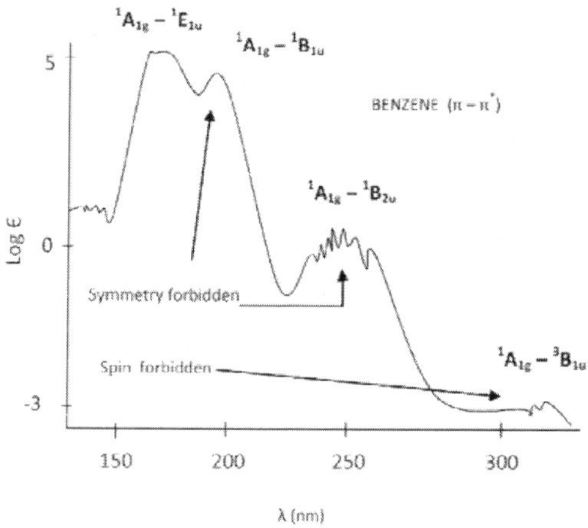

Figure:

Organic Molecule Spectra

From the example of benzene, we have investigated the characteristic pi to pi* transitions for aromatic compounds. Now we can move to other organic molecules, which involves n to pi* as well as pi to pi*. Two examples are given below:

The highest energy transition for both of these molecules has an intensity around 10,000 cm^{-1} and the second band has an intensity of approximately 100 cm^{-1}. In the case of formaldehyde, the n to pi* transition is forbidden by symmetry where as the pi to pi* is allowed. The opposite is true for $As(Ph)_3$ and the difference in molar absorptivity is evidence of this.

n to pi* transitions: These transitions involve moving an electron from a nonbonding electron pair to a antibonding pi* orbital. They tend to have molar absorbtivities less than 2000 and undergo a blue shift with solvent interactions (a shift to higher energy and shorter wavelengths). This is because the lone pair interacts with the solvent, especially a polar one, such that the solvent aligns itself with the ground state. When the excited state emerges, the solvent molecules do not have time to rearrange in order to stabilise the excited state.

This causes a lowering of energy of the ground state and not the excited state. Because of this, the energy of the transition increases, hence the "blue shift".

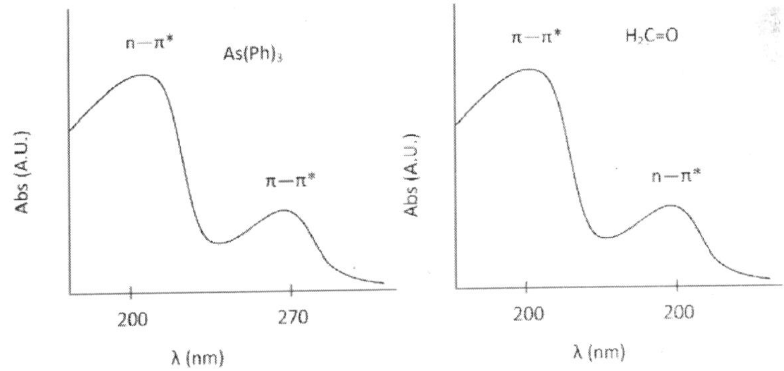

Figure:

pi to pi* transitions: These transitions involve moving an electron from a bonding pi orbital to an antibonding pi* orbital. They tend to have molar absorptivities on the order of 10,000 and undergo a red shift with solvent interactions (a shift to lower energy and longer wavelengths). This could either be due to a raising of the ground state energy or lowering of the excited state energy. If the excited state is polar, then it will be solvent stabilised, thus lowering its energy and the energy of the transition.

Inorganic Molecule Spectra

Speaking of transition probabilities in organic molecules is a good seq way into interpreting the spectra of inorganic molecules. Three types of transitions are important to consider are Metal to Ligand Charge Transfer (MLCT), Ligand to Metal Charge Transfer (LMCT), and d-d transitions. To understand the differences of these transitions we must investigate where these transitions originate. To do this, we must define the difference between pi accepting and pi donating ligands:

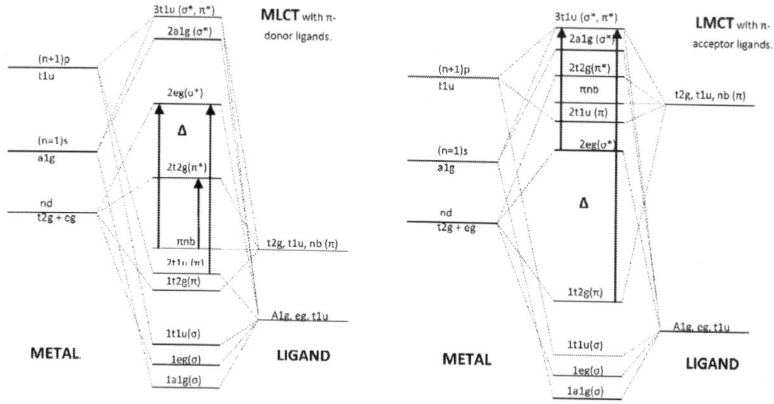

Figure:

d-d Transitions

From these two molecular orbital energy diagrams for transition metals, we see that the pi donor ligands lie lower in energy than the pi acceptor ligands. According to the spectral chemical series, one can determine whether a ligand will behave as a pi accepting or pi donating. When the ligand is more pi donating, its own orbitals are lower in energy than the t2g metal orbitals forcing the frontier orbitals to involve an antibonding pi* (for t2g) and an antibonding sigma* (for eg). This is in contrast to the pi accepting ligands which involve a bonding pi (t2g) and an antibonding sigma* (eg). Because of this, the d-d transition (denoted above by delta) for the pi acceptor ligand complex is larger than the pi donor ligand. In the spectra, we would see the d-d transitions of pi acceptor ligands to be of a higher frequency than the pi donor ligands. In general though, these transitions appear as weakly intense on the spectrum because they are Laporte forbidden. Due to vibronic coupling; however, they are weakly allowed and because of their relatively low energy of transition, they can emit visible light upon relaxation which is why many transition metal complexes are brightly coloured. The molar extinction coefficients for these transition hover around 100.

LMCT Transitions

At an even higher energy are the LMCT which involve pi donor ligands around the metal. These transitions arise because of the low-lying energy of the ligand orbitals. Therefore, we can consider this as a transition from orbitals that are ligand in character to orbitals that are more metal in character, hence the name, Ligand to Metal

Charge Transfer. The electron travels from a bonding pi or non-bonding pi orbital into a sigma* orbital. These transitions are very strong and appear very intensely in the absorbance spectrum. The molar extinction coefficients for these transitions are around 10^4. Examples of pi donor ligands are as follows: F^-, Cl^-, Br-, I^-, H_2O, OH^-, RS^-, S_2^-, NCS^-, NCO^-,...

MLCT Transitions

The somewhat less common MLCT has the same intensity and energy of the LMCT as they involve the transition of an electron from the t2g (pi) and the eg (sigma*) to the t1u (pi*/sigma*). These transitions arise from pi acceptor ligands and metals that are willing to donate electrons into the orbitals of Ligand character. This is the reason that they are less frequent since metals commonly accept electrons rather than donate them. All the same, both types of Charge Transfer bands are more intense than d-d bands since they are not Laporte Rule forbidden. Examples of pi accepting ligands are as follows: CO, NO, CN^-, N_2, bipy, phen, RNC, $C_5H_5^-$, C=C double bonds, C=C triple bonds,...

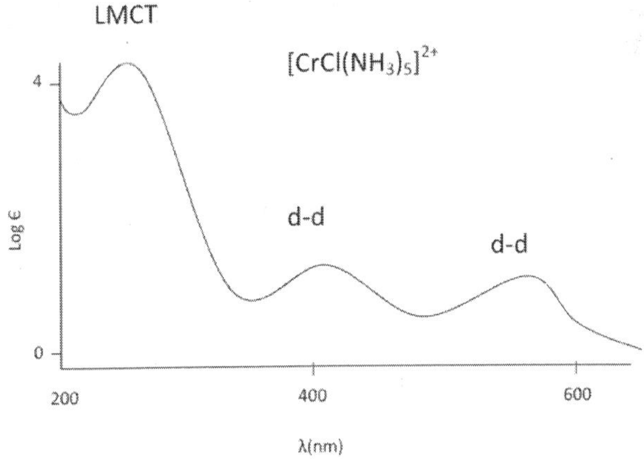

Figure:

From this spectra of an octahedral Chromium complex, we see that the d-d transitions are far weaker than the LMCT. Since Chlorine is a pi donor ligand in this example, we can label the CT band as LMCT since we know the electron is transitioning from a MO of ligand character to a MO of metal character. The Laporte forbidden (symmetry forbidden) d-d transitions are shown as less intense since they are only allowed via vibronic coupling. In addition, the d-d transitions are lower in energy than the CT band because of the smaller energy gap between the t2g and eg in octahedral complexes (or eg to t2g in tetrahedral complexes) than the energy gap between the ground and excited states of the charge transfer band.

These transitions abide by the same selection rules that organic molecules follow: spin selection and symmetry arguments. The Tanabe and Sugano diagrams for transition metal complexes can be a guide for determining which transitions are seen in the spectrum. We will use the [CrCl(NH3)5]2+ ion as an example for determining the types of transitions that are spin allowed. To do this we look up the Tanabe and Sugano diagrams for Octahedral fields. Since Cr in the complex has three electrons, it is a d3 and so we find the diagram that corresponds to d3 metals:

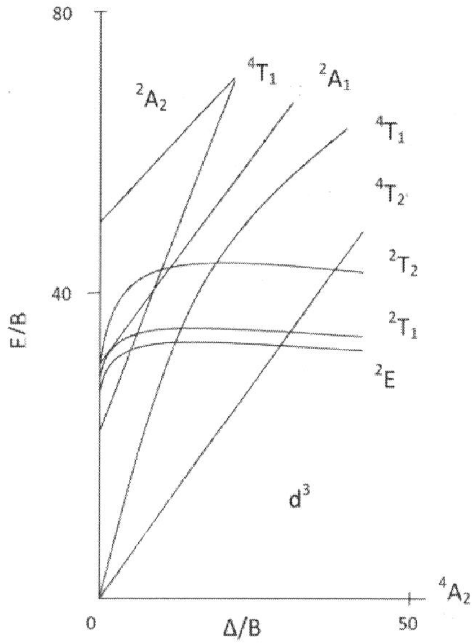

Figure:

Based on the TS diagram on the left, and the information we have already learned, can you predict which transition will be spin allowed and which ones will be forbidden? From the diagram we see that the ground state is a $^4A_2$. This is because of the three unpaired electrons which make M=2S+1= 4. The A comes from the fact that there is only one combination of electrons possible. With a spin multiplicity of 4, by the spin selection rules, we can only expect intense transitions between the ground state $^4A_2$ and $^4T_2$, $^4T_1$, and the other $^4T_1$ excited state. The other transitions are spin forbidden. Therefore, we would expect to see three d-d transitions on the absorption spectra.

For us to visualize this, we can draw these transitions in order of increasing energy and then plot the spectrum as we would expect it for only the d-d transitions in a d^3 octahedral complex:

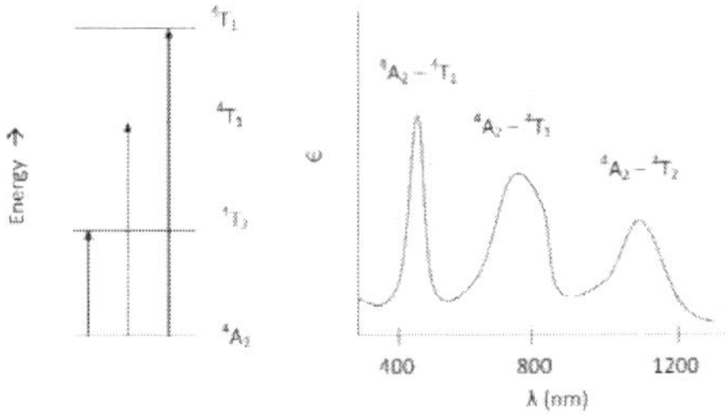

Figure:

From three spin allowed transitions, we would expect to see three d-d bands appear on the spectrum. In addition to these of course, the LMCT band will appear as well.

Fluorescence

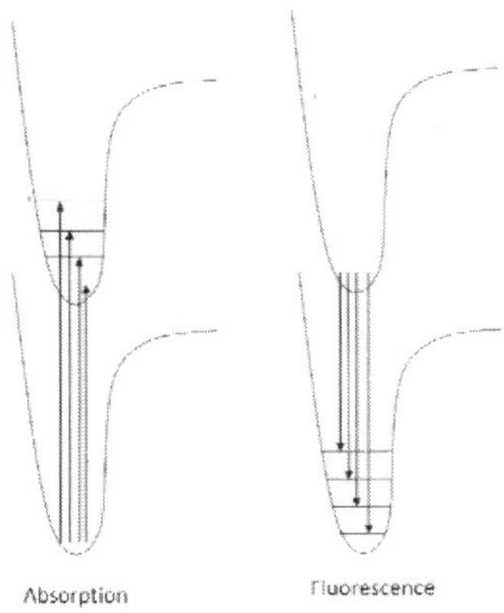

Figure:

Now that we have discussed the nature of absorption involving an electron absorbing photon energy to be excited to a higher energy level, now we can discuss what happens to that excited electron. Due to its higher potential energy, the electron will relax back to its initial ground state, and in the process, emit electromagnetic radiation. The energy gap between the excited state and the state to which the electron falls determines the wavelength of light that will be emitted.

This process is called fluorescence. Generally, the wavelengths of fluorescence are longer than absorbance, can you explain why? Given the following diagram, one can see that vibrational relaxation occurs in the excited electronic state such that the electronic relaxation occurs from the ground vibrational state of the excited electronic state. This causes lower energy electronic relaxations than the previous energy of absorption

Here we see that the absorption transitions by default involve a greater energy change than the emission transitions. Due to vibrational relaxation in the excited state, the electron tends to relax only from the v'=0 ground state vibrational level. This gives emission transitions of lower energy and consequently, longer wavelength than absorption. When obtaining fluorescence, we have to block out the transmitted light and only focus on the light being emitted from the sample, so the detector is usually 90 degrees from the incident light. Because of this emission spectra are generally obtained separately from the absorption spectra; however, they can be plotted on the same graph as shown.

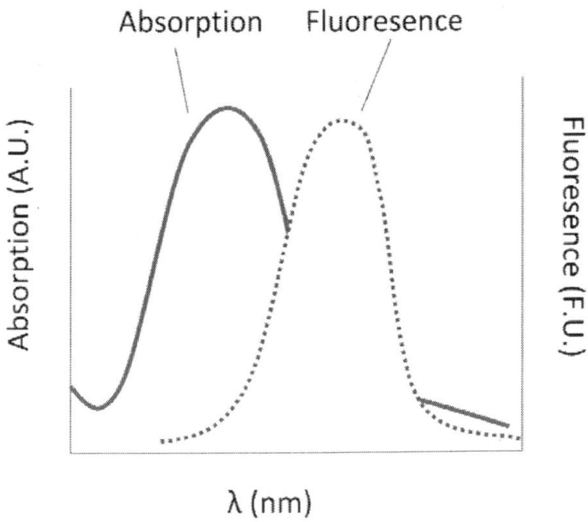

Figure:

Generally separated by ~10nm, the fluorescence peak follows the absorption peak according to the spectrum. With that, we conclude our discussion of electronic spectroscopy interpretation. Refer to outside links and references for additional information.

Electronic Spectra of Molecules

Absorption or emission of ultraviolet or visible light by a molecule depends on electron transitions between molecular orbital energy levels, just as absorption or emission of electromagnetic radiation by an atom is determined by electron transitions between different energy levels in the atom and the DEs for those transitions. Molecular spectra follow rules analogous to the rules for atomic spectra: energy is absorbed only when the amount of energy provided matches the difference in energy, DE, of 2 energy levels.

When an electron goes from a higher to a lower energy state, a photon of definite wavelength and frequency is emitted. Every atom or molecule has a characteristic electronic spectrum depending on its characteristic DEs. Because of a molecule's greater complexity, we can often construct a molecule that will give a particular spectrum, rather than having to just accept the spectra available as we do with atoms. This possibility arises because of the interdependence of molecular orbital energy level values for the molecule, molecular shape, bonding, and distribution of electron density within the molecule. Energy transitions in phenolphthalein indicator, plant pigments, and isolated pi systems will be examined to illustrate this.

Molecular orbital theory provides a model for the way electromagnetic radiation interacts with molecules. For example, an electron in the pi bonding molecular orbital (MO) of an alkene can be excited to a pi antibonding MO. This is described as a Π to Π^* transition.

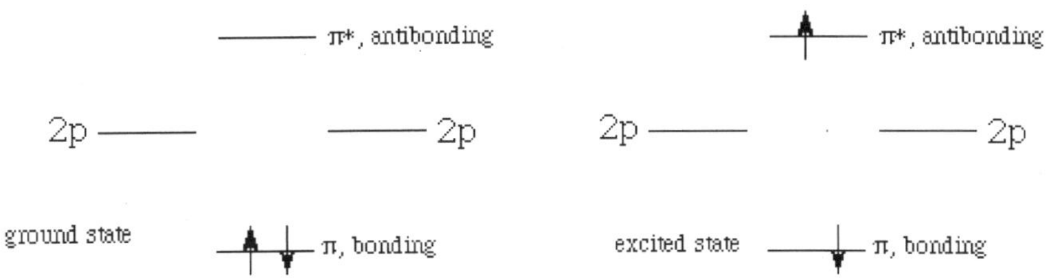

For an isolated pi bond the energy separation, DE, between the pi bonding and pi antibonding MO's is large; ultraviolet light with its large energy and short wavelength is needed to excite the pi electron. Molecular orbital theory predicts that the energy difference, DE, between levels will decrease if the double bond is conjugated with another double bond. (Conjugated double bonds have one single bond separating them.

They are coplanar and pi electrons can move through out the pi system.) Cogjugation exists when series of alternating double and single bond. This means that the molecule has a single bond between the two double bonds.

The term "conjugated" is used in chemistry to refer to a series of alternating single and double bonds. The predicted decrease in DE for conjugated structures is also observed in experiments. Here is a general rule that describes the effect of double bonds conjugation on the energy absorbed by the pi system. *The greater the number of conjugated multiple bonds in a compound, the longer the wavelength of the light that the compound will absorb.* This translates into a smaller energy requirement for the transition from the highest occupied MO to the lowest unoccupied MO. This idea is illustrated schematically in the following diagram for alkenes where the transition is for

$$\prod to \prod^{*}$$

As the conjugation increases, ¼ electrons are delocalised over more space and the system is more stable.

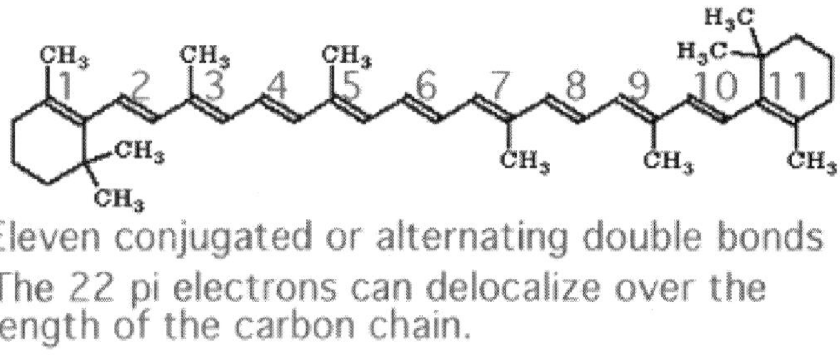

Figure: There are 11 double bonds in the alternating double bond-single bond system in b-carotene. The double bonds are essentially coplanar and the ¼ electrons are delocalised over a 22 carbon chain.

The wavelength absorbed by b-carotene is in the blue region of the visible spectrum. When we look at b-carotene, no "blue wavelengths" reach our eyes because the carotene molecule is absorbing these. The rest of the wavelengths in the visible region (minus the blue) reach our eyes and carotene looks orange to us. Remember, the colour of light absorbed by the molecule is not the colour we see. It is the one we don't see!

Observed Colour of Compound	*Colour of Light Absorbed*	*Approximate Wavelength of Light Absorbed*
Green	Red	700 nm
Blue-green	Orange-red	600 nm
Violet	Yellow	550 nm
Red-violet	Yellow-green	530 nm
Red	Blue-green	500 nm
Orange	Blue	450 nm
Yellow	Violet	400 nm

Absorption or emission of ultraviolet or visible light by a molecule depends on electron transitions between molecular orbital energy levels, just as absorption or emission of electromagnetic radiation by an atom is determined by electron transitions between different energy levels in the atom and the DEs for those transitions. Molecular spectra follow rules analogous to the rules for atomic spectra: energy is absorbed only when the amount of energy provided matches the difference in energy, DE, of 2 energy levels. When an electron goes from a higher to a lower energy state, a photon of definite wavelength and frequency is emitted. Every atom or molecule has a characteristic electronic spectrum depending on its characteristic DEs. Because of a molecule's greater complexity, we can often construct a molecule that will give a particular spectrum, rather than having to just accept the spectra available as we do with atoms. This possibility arises because of the interdependence of molecular orbital energy level values for the molecule, molecular shape, bonding, and distribution of electron density within the molecule. Energy transitions in phenolphthalein indicator, plant pigments, and isolated pi systems will be examined to illustrate this.

Phenolphthalein is another example of the effect of molecular shape on delocalisation of pi electrons. Phenolphthalein molecule is colourless in acid solution and its pi system is nonplanar. The electrons in the double bonds are "localised. The excitation energy is large and in the ultraviolet range. The molecule loses protons in basic solutions and forms a minus 2 anion. This negative ion is flat and the pi system is conjugated. In the planar ion the pi electrons in the eleven double bonds can delocalise over the 3 rings, the COO^{-1} carboxyl group and the oxygens. The anion imparts a colour to solutions because the energy separation between the pi and pi antibonding levels is smaller and matches light in the visible range.

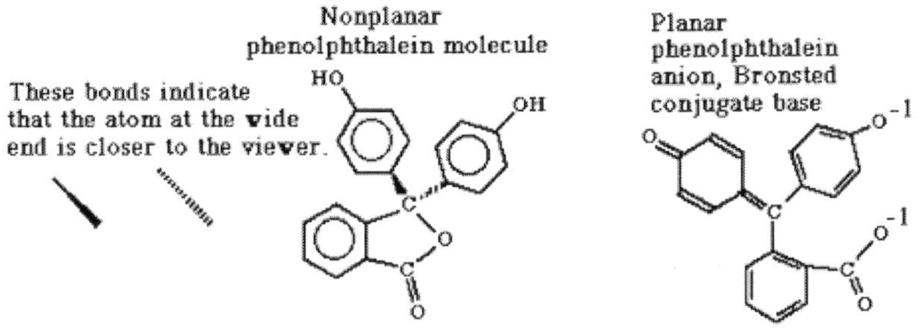

Figure: In the first part of this experiment you will observe this difference in colour for phenolphthalein and its anion.

In the second part of this experiment you will extract chlorophyll from green leaves and then use chromatography to separate *chlorophyll a* from *chlorophyll b*. You may also see the separation of carotene and other plant pigments. You will be asked to explain the different colours of the 2 chlorophylls by examining their structures and identifying the type of change in molecular orbital energy levels caused by the difference in structure. Paper chromatography is a separation technique that anyone who ever spilled coffee or tea on a piece of paper has seen. The solvent wets the paper, creeps along carrying solutes along with it. The different dissolved materials move with the solvent, but at different rates because the paper attracts the solutes differently. You will use this behaviour in a systematic way to separate pigments in chlorophyll. You will make an extract of leaf pigments by soaking plant leaves in a mixture of cold acetone and ethanol. The extract appears to be green, but other pigment colours may be masked by the strong green tint. You will use paper chromatography to separate any pigments in the extract. The separation occurs because the pigments have different solubilities in the mobile solvent and different absorption affinities (attractions) to the chromatography paper. The separation process will give a chromatogram.

Equipment and Supplies	
Al foil	1-3 g fresh spinach leaves
2 beakers, 100 mL	ice
graduated cylinder, 10 mL	phenolphthalein indicator solution
medicine cup	acetone
3 Beral type pipets	vinegar
erlenmeyer flask, 50 mL	ethanol
Beral type pipet with narrow stem	5M NaOH
glass jar about 7 inches tall (about 1 quart)	petroleum ether
2 pencils	chromatography paper
string	plastic wrap
scissors	2 rubber bands
	tape

Procedure

Please read the entire procedure before starting the experiment.

Wear your goggles at all times while doing the experiment!

Gather your equipment and chemicals. Cover your work area with Al foil or waxed paper to catch spills and spatters.

Preparation for Extracting Chlorophyll: Do this first, then do the phenolphthalein part of the experiment. Finally, finish up with the chlorophyll chromatography. Collect 1 to 3 grams of fresh spinach leaves. If you can't get spinach, other fresh green leaves (turnip greens, broccoli leaves) will work, but your chromatogram may be slightly different from the one described later in this experiment. Any dark green leaves should allow you to extract the 2 chlorophyll pigments. Make an ice bath by placing a little cold water and some ice in a beaker. Remove the containers of acetone and ethanol from your kit. Do not open yet. Set the 2 containers in the ice bath so the acetone and the ethanol get cold. Do not let the water get into the acetone or ethanol containers.

Phenolphthalein Molecule and Anion Colours: Pour about 2 mL of water into one of the medicine cups from your kit. The reference marks on the cup are adequate; you do not need to measure the water with your graduated cylinder.

Add one or 2 drops of the phenolphthalein indicator solution to the water in the medicine cup. You can do this with one of your Beral type pipets if there is no "built in" dropper in your phenolphthalein container.

Use a different Beral type pipet to add 2 drops of vinegar to the medicine cup. Stir with your stirring rod. Note and record the colour of the solution in the medicine cup.

USE CARE WHEN POURING THE NaOH. IT IS CAUSTIC.

If the 5.0 M NaOH solution gets into your eyes, rinse the eyeball for 15 minutes in warm running water (force your eyelids apart with your fingers to keep your eye open and the eyeball exposed to the running water). Contact your physician for advice if your eye continues to hurt or "burn".

Use a third Beral type pipet to add 2 or 3 drops of 5 M NaOH to the contents of the medicine cup. Stir with your stirring rod (no need to wash the rod between steps). Note and record the colour of the mixture in your cup. Discard the solution in the medicine cup by pouring it into the drain followed by cold water.

Preparation of Pigment Extract: Work in a well ventilated area away from any open flames or hot surfaces (stove, space heater, water heater, etc.). Acetone and ethanol convert to vapour very easily, especially the acetone. Both the liquid and the vapour are flammable!

Label your erlenmeyer flask “pigment extract mixture”.

Cut the leaves into tiny pieces using your scissors. The smaller the pieces, the better the contact between pieces of leaf material and the extraction solvent.

Place the small leaf pieces in the erlenmeyer flask and add 5 mL of cold ethanol.

Then add 5 mL of cold acetone. There is no need to wash your graduated cylinder between the ethanol and acetone.

Quickly cover the erlenmeyer flask with a piece of plastic. Hold the plastic wrap snugly against the neck of the flask with a rubber band. This helps keep the solvent from evaporating while the pigments are being extracted from the leaves.

Dump the water out of the beaker. Keep the ice. Set the erlenmeyer flask in the ice in the beaker. Store both in a cool, well ventilated area for about 24 hours. Do not put the beaker and erlenmeyer in your refrigerator. Store away from food, hot surfaces, and open flames. The solvent will extract (dissolve) pigments from the leaf material. Then you can do the chromatography to separate the different pigments.

Chromatography

Assemble a chromatography chamber. You need a glass jar that is about 7 inches tall, a piece of plastic wrap to cover the top of the jar, and a large rubber band to go around the neck of the jar to hold the plastic wrap tightly in place. A recycled jelly jar, instant coffee jar, or just about any quart jar will do. Obtain a piece of chromatography paper that measures about 3 cm by 15 cm. Handle only the edges. Try not to touch the surface. Oils from your fingers can alter the absorptive properties of the paper and the separation may not happen normally. Use a ruler and a pencil to draw a line one centimetre from the bottom 3 cm edge of the paper. DO NOT USE INK! Ink will dissolve in the solvent and ruin the separation process. This is the line where you will place a spot of the extract. Lightly mark the edges at one-centimetre distances along the edges of the paper, lengthwise. This helps you keep track of the leading edge of your solvent (the solvent front) as it "climbs" the strip of paper. Use the tip of your scissors to poke a small hole about 1/2 cm from the top 3 cm edge of the paper. Insert a piece of string or thread through the hole. Tape the ends of the string to the middle of a pencil or pen. The paper should hang freely from the pencil when you hold up the pencil. Hang the paper in the centre of your glass jar by placing the pencil across the mouth of the jar. Rotate the pencil to wind the string around it until the bottom of the paper just clears the bottom of the glass jar. The paper should not touch any part of the jar, not even the bottom.

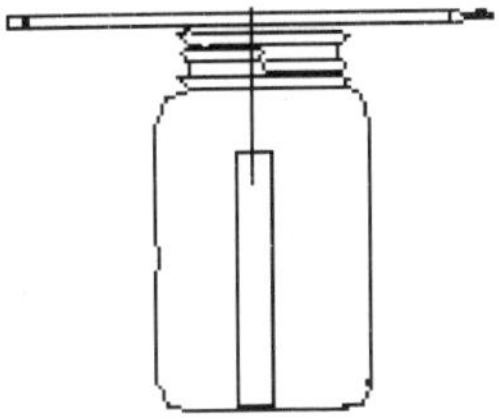

Figure: Tape the wound-up string in place on the pencil so it can't unwind when you let go of it. Remove paper from jar by lifting the pencil. Apply the "pigment extract mixture" as directed below.

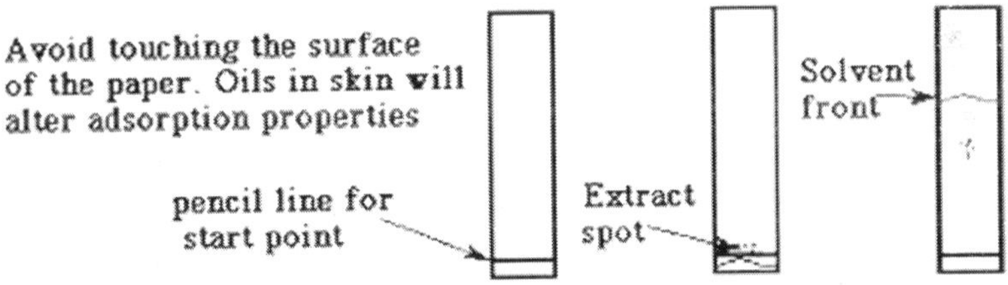

Figure:

This animation illustrates what happens in the separation of a two component mixture. The solvent front is marked by the upper black line. There is a green component that moves faster than the yellow component in the mixture. The Rf value for the green is closer to "1" than the Rf value for the yellow component.

Take the plastic wrap off the beaker of "pigment extract mixture". Use a Beral type pipet with a very narrow stem to draw up a very small quantity of pigment extract.

Place one drop of the pigment extract on or just slightly above the pencil line on the chromatography paper. The pigment extract spot should be about 3 to 5 mm in diametre which is the size of this spot. Repeat this procedure to apply 4 or 5 more drops directly on top of the first drop. ALLOW THE PIGMENT TO DRY BETWEEN APPLICATIONS ! If you apply a large amount at one time or if the spot is still wet from the previous drop, the spot will spread over an unmanageable area and give poor separation results. You need to have several drops of pigment on the paper to make the colours of minor parts of the mixture visible in your finished chromatogram.

Place the plastic wrap cover back on the erlenmeyer flask of pigment extract. You may need to repeat the chromatography separation and need this extract for the duplicate trial. Prepare a mixture of 1 mL acetone and 9 mL petroleum ether. This is the solvent for the chromatography separation.

Place enough acetone-petroleum ether solvent in your "chamber" to give a depth of 1/2 cm in the bottom of your glass jar. The liquid level must be lower than the pencil start line across the paper.

Place the "loaded" chromatography paper into the chamber. Lower the pencil until the bottom of the paper just touches the solvent in the bottom of the jar. You want the bottom of the paper to get wet with solvent. You do *not* want the spot of pigment extract to touch the solvent at this time.

Cover the top of the chamber with plastic wrap; use a rubber band to hold the wrap in place.

Watch the solvent climb the paper. Gauge the speed of movement of the solvent front. The progress of the solvent will be quick at first. The pigment extract spot will start to smear out as the solvent carries the pigments away from the start line and the chromatogram "develops".

Allow the chromatogram to develop for 10 minutes, or until the solvent front comes within 1 cm of the top edge of the paper. Remove the paper from the chamber by lifting the pencil.

If you let the chromatogram develop too long, the solvent front will actually come to a stop. Solvent molecules will be evaporating from the wet paper as fast as they are "climbing" onto the paper at the bottom, so the front will stop rising. However, solvent molecules will still be moving up the paper, washing the pigment along. The different components of the pigment mixture will start to over-take one another and "pile up", undoing the separation you achieved. Allow the paper to dry. Use pencils to sketch the outlines of the colours on the paper. Begin with the colour closest to the start line and label the different colour regions. The first pigment spot should be yellow-green; this is *chlorophyll b*. The blue-green pigment is *chlorophyll a*. The darker orange spot near the top is *carotene*. Xanthophylls and carotenes are members of the carotenoid class of pigments. The chromatogram will fade on exposure to light because of light induced oxidation reactions. If you want to preserve it, store it in a dark place.

Vibrational and Electronic Energy Levels of Polyatomic Transient Molecules

Most chemical processes—including not only laboratory and industrial chemical syntheses but also those which occur in flames, propellant systems, the initiation of energetic materials, atmospheric pollution, chemical vapour deposition, and plasma processing—consist of a complicated sequence of interrelated reactions in which neutral and charged molecular fragments play essential roles. Although these fragments are present in only very small concentration, they are highly chemically reactive. If a specific molecular fragment is removed from the system, as by introducing a scavenger molecule, the reactions in which that fragment participates stop. Other parts of the overall process continue, resulting in very significant changes in product distribution and yield.

In the early studies of complex chemical processes, it was necessary to postulate mechanisms involving such transient intermediates, present in concentrations too small for direct detection. Conventional end product analysis aids in the selection of suitable mechanisms, but generally does not yield a complete description of the system. Consequently, the improvement of industrial chemical processes often is achieved by semiempirical experimentation. The determination of the detailed chemical mechanism would, in turn, permit the development of rational strategies for removing undesired products and enhancing the yield of the desired species.

In recent years, there has been great progress in the development of techniques suitable for monitoring chemical reaction intermediates. Molecular spectroscopy is especially well suited to this task. Optical detection can be used not only for gas-phase measurements, but also for studies of processes that occur on surfaces or in the condensed phase. It also permits remote sensing, an important advantage. A wide variety of recently developed laser-based spectroscopic detection schemes are not only highly sensitive but also space and time specific. Although the development of spectroscopy-based diagnostics for chemical reaction systems is in its infancy, already the laboratory application of sophisticated sampling and observation techniques has yielded a wealth of vibrational and electronic spectral data for reaction intermediates.

Earlier Data Compilations

For many years, the most important source of vibrational and electronic energy level data for small polyatomic reaction intermediates was the compilation of spectroscopic data for small polyatomic molecules (3-12 atoms) given by Herzberg. To meet the need for an updated, critically evaluated compilation, a series of publications have appeared in the Journal of Physical and Chemical Reference Data, culminating in the publication in 1994 of a monograph (Vibrational and Electronic Energy Levels of Polyatomic Transient Molecules) which presented evaluated spectral data for more than 1550 small polyatomic transient molecules, defined as species that have a lifetime of less than a few minutes in the pressure range (typically 0.1 to 1.0 Torr) encountered in their production. Vibrational fundamentals in the ground and excited electronic states and radiative lifetimes were included. To aid in spectral identification, the principal rotational constants were also given to three decimal places. These tables provide the basis for this database, designed to supplement the published compilation by providing a capability for rapid searches by molecule or wavenumber.

Recent Progress

The rapid growth in the scientific literature concerned with the spectroscopic study of transient molecules and with their detection in chemical reaction systems continues. Since the October 1993 cutoff in the data evaluation for the monograph, substantive new spectroscopic data have been published for many species included in it, and the first data have become available for almost 2000 other molecules. There has been especially great progress in the spectroscopic characterisation of transient species produced by the reaction of metal atoms with oxygen and other small molecules. Despite this rapid progress, many gaps remain in our knowledge of the energy levels of the species represented, and many new and potentially important transient molecules are still to be discovered. This compilation attempts to provide a comprehensive, critically evaluated summary of vibrational and electronic energy level data for small polyatomic transient molecules, in order to support further research and new technologies such as those of plasma processing and chemical vapour deposition.

Scope of Database

A critical evaluation and summary of experimental vibrational and electronic energy level data for neutral and ionic transient molecules and high temperature species possessing from three to sixteen atoms is presented. By early 2003, evaluated data were available for approximately 3500 molecules, and the published version encompassed non only the original monograph, but also two supplements. The Chemistry WebBook database brings these results together in an ongoing effort to provide evaluated data to the scientific community. Although radiative lifetimes and principal rotational constants are not presently included in this version of the database, references to the original literature concerned with then are given.

Stable Molecules and High-Temperature Species

Data have been selectively included for some molecules which are important in environmental and industrial chemical reaction systems but which can be studied only with difficulty using conventional sampling techniques because of the ease with which they decompose, rearrange, or polymerize. Also included are data derived from spectra of many high-temperature species, such as metal oxides, studied in molecular beams and in rare-gas matrices. Unfortunately, with these few exceptions it is not possible to include data for stable molecules. However, the spectra of many of these species are relatively well established, and sources of data such as the tables of Herzberg remain extremely useful. *In obtaining spectral identifications with the help of the present data, it is crucial that the possible contribution of the absorptions or emissions by a stable molecule also be considered.*

Types of Measurement

Ground-State Vibrational Spectra

Gas-Phase Measurements: Studies in the gas phase offer the potential for the most precise, detailed measurements. Because of the high chemical reactivity of transient molecules, it is difficult to obtain gas-phase survey infrared spectra of them. The well known advantages of Fourier transform infrared measurements, coupled with sophisticated digital data handling procedures, have permitted the acquisition of gas-phase survey spectra for a number of transient molecules.

Vibrational Frequencies from Rydberg Transitions

Although vibrational frequencies of ground-state molecular ions have frequently been estimated from structure in Rydberg transitions of the parent neutral species, in this compilation vibrational frequencies of the ions are not inferred from these Rydberg transitions. Many of these transitions have residual valence character, resulting in significant variations in vibrational frequencies from one Rydberg state to another.

Matrix Isolation Measurements

As in the earlier compilations, spectral data obtained for molecules trapped in dilute solid solution, with the solvent a rare gas or a small covalent molecule, are included. The application of this sampling technique, known as matrix isolation, for the stabilisation and spectroscopic study of uncharged reaction intermediates has recently been reviewed. Because nitrogen and the rare gases are transparent through the entire infrared spectral region, matrix isolation measurements provide a potentially valuable survey tool. In these matrices, infrared absorptions are typically sharp, with half band widths between 0.1 and 1 cm^{-1}. Rotational structure is, with few exceptions, quenched. Multiple trapping sites occur, often resulting in the appearance of several absorption maxima—usually one or two of which predominate—over a range of a few cm^{-1}.

Matrix Shifts for Uncharged Covalently Bonded Molecules

Matrix shifts for covalently bonded molecules trapped in solid neon or argon often are quite small. A comparison of the positions of the ground-state vibrational fundamentals of over two hundred diatomic molecules observed in the gas phase and in nitrogen and rare-gas matrices has shown that, typically, the smallest matrix shift occurs for neon matrix observations, with successively greater matrix shifts for the heavier rare gases and for nitrogen. Except for very weakly bonded molecules and for the alkali metal and Group IIIa halides, matrix shifts of most diatomic molecules isolated in solid argon are smaller than 2%. Similar conclusions resulted from a comparison of neon- and argon-matrix shifts for the ground-state vibrational fundamentals of larger molecules.

The generalization that matrix interactions are minimal for neon and that they increase as the mass of the rare gas is increased and become even more important for nitrogen and most other small molecule matrices is supported both by experimental observations on larger molecules and by ab initio calculations for the weakly bonded CaH_2 and CaF_2 molecules complexed with the rare gases and with nitrogen. For neon matrices, the maximum in the distribution lies near 0.0%, and for argon matrices, near 0.2%. For both neon and argon matrices, fewer than one-tenth of the matrix shifts are greater than 1%.

Data are beginning to appear for molecules trapped in a hydrogen matrix. Insufficient information is available for generalization on the magnitude of matrix shifts in this medium. For the few species heretofore studied, including several transient molecules present in this compilation, the matrix shifts have been comparable to those in a neon matrix.

Many other matrix materials have also been employed for spectroscopic studies. However, complications due to reaction or to relatively strong interaction (e.g., hydrogen bonding) of the transient molecule with the matrix frequently occur. Therefore, observations in such media as solid hydrocarbons and aqueous solutions and studies of condensed reaction products without an inert carrier have been excluded.

Matrix Shifts for Molecular Ions

For molecular ions, neon is the matrix of choice. Polarization and charge-transfer interactions become successively more important for molecules isolated in the heavier rare gases. Charge delocalisation sometimes also occurs for ionic species trapped in the rare gases. The anomalously large matrix shift for ν_3 of $ClHCl^-$ may be attributed to this phenomenon. Data for small cation species trapped in solid neon are consistent with the matrix shift generalizations given above. Only twelve comparisons are available for molecular cations observed both in the gas phase and trapped in solid argon.

The absolute values of five of the observed matrix shifts are greater than 1%. For several other vibrations of molecular ions which have been observed in both neon and argon matrices but not in the gas phase, there are deviations greater than 1% between the neon- and argon-matrix frequencies. Very few comparisons are possible for molecular anions. A number of these species have been generated in rare-gas (usually argon) matrices by charge transfer between a precursor molecule and an alkali metal atom. Recent studies of such species as CO_2^- and SO_2^- generated instead by photoionization and/or Penning ionization and trapped in solid neon indicate that shifts on the order of 50 cm^{-1} may be attributed to the relatively strong interaction of the anion with the nearby alkali metal cation. On the other hand, when the uncharged molecule has a relatively large electron affinity, as is true for C_2 and for NO_2, charge transfer occurs at a relatively great separation, and a substantial fraction of the anion population may be trapped in sites in which interaction with the alkali metal cation is minimal.

Matrix Shifts for Ionic Bonds

Matrix shifts for vibrations associated with ionic bonds are often considerably larger than those associated with uncharged molecules or with intramolecular vibrations of molecular ions. Criteria for inclusion of data for species that include an ionic bond are exemplified by the selection process for the heavy-metal oxides. Often the stable dioxide structures include an $M^+O_2^-$ species with significant covalent bond character for the attachment of M^+, evidenced by a substantial shift in the O_2^- stretching fundamental as M^+ is varied. Such species are included in the compilation. On the other hand, there is little evidence for substantial metal-atom participation in the vibrations characteristic of the O_3^- moiety of $M^+O_3^-$. Accordingly, spectral data are given for O_3^-, but not for $M^+O_3^-$.

Electronic Spectra

Gas-Phase Electronic Spectra

Gas-phase studies of the electronic spectra of transient molecules were for many years much more readily conducted than were studies of ground-state vibrational spectra.

The concentration of transient molecules in flames, chemiluminescent reactions, or various types of discharge may be sufficiently high for spectroscopic study. The photographic plate provides a cumulative detector for visible and ultraviolet radiation. Much of the electronic spectral data summarised in these tables was obtained using conventional gas-phase ultraviolet absorption or emission spectroscopy, which affords the potential for both a broad spectral survey and very high resolution.

Flash Photolysis

Flash photolysis permits the observation of relatively high concentrations of transient species at short time intervals after the flash. Because the products are generally formed with much less internal energy than is typical of systems with detectable emission spectra, the absorption spectra obtained in flash photolysis studies are more readily analysed. Furthermore, the time-resolved detection used in flash photolysis studies provides information on the rates of formation and disappearance of transient molecules in the system.

Laser-Based Techniques

More recently, a wide variety of laser-based techniques have also been used for electronic spectral observations, often with exceptionally high detection sensitivity. In recent years, very high resolutions have been achieved in laser-based spectroscopic measurements. Since a given laser is tunable over a relatively limited spectral region, laser studies of transient molecules are greatly aided by the availability of survey spectra obtained using other techniques. The spatial configuration of the laser beam makes it an extremely powerful tool for studies of the energy levels of molecules in molecular beams. If the molecule of interest is present in a supersonic molecular beam, excited rotational and vibrational energy levels can be very effectively depopulated, and the absorption spectrum of the molecule is greatly simplified. The spatial configuration of the laser beam also makes it amenable to the development of probes for chemical reaction intermediates not only in the laboratory but also in the environment and in industrial processes.

Laser studies may be broadly classified according to whether the interaction of the molecule with the laser beam(s) is followed by photon or mass detection. Photon-based observations are amenable to remote sensing applications. Because pulsed lasers offer an exceptionally wide range of time specificity, they are very useful for determining radiative lifetimes and rates of elementary chemical reactions. The coupling of lasers with mass detection can lead to the identification of transient molecules for which survey spectra are not available. Among the mass-based detection schemes are photofragment spectroscopy and resonance-enhanced multiphoton ionization (REMPI).

REMPI provides a powerful tool for mapping the Rydberg transitions of transient molecules. Whereas laser-excited fluorescence measurements depend upon the presence of electronic energy levels which decay by photon emission, all molecules possess Rydberg energy levels. REMPI measurements depend on multiphoton excitation into a suitable electronic energy level, most often one of Rydberg character. The selection rules may permit excitation of levels which are not accessible by one-photon excitation from the ground state. The range of tunability of the laser is multiplied by the number of photons required for the excitation of the Rydberg level, significantly broadening the spectral regin which can be probed with a given laser. When the parent molecule is a free radical, almost all of the mass signal is generally found to arise from the parent cation, with very little fragmentation.

Photoelectron Spectroscopy

Much valuable information on the energy levels of molecular cations has been obtained from photoelectron spectroscopy. These tables include selective coverage of the voluminous literature on photoelectron spectroscopic measurements. The number of stable molecules that possess more than six atoms for which photoelectron spectra have been reported is too great to permit the inclusion of low-to-moderate resolution photoelectron spectral data for molecular cations with more than six atoms. Those who need such data for larger molecules may find the reviews by Turner et al. Rabalais, and Kimura et al. helpful.

Several criteria are important in determining whether a given reference should be included in the present work. The first of these is resolution. In the few instances in which high resolution photoelectron data are available, these data are heavily weighted. Where direct spectroscopic observation is possible, the measurements generally are of considerably higher precision than are the photoelectron data, which are then omitted from the tables. A second criterion is the availability of adiabatic ionization potentials. In order to obtain information on the positions of electronic transitions from photoelectron spectral data, it is necessary to subtract the first ionization potential from the energy of the photoelectron band. Where there is little change in the molecular geometry in the transition, the difference between the vertical ionization potentials gives a reasonable approximation to the position of the electronic transition. However, this is not the general case. Therefore, priority is given to papers that include adiabatic ionization potentials.

For most photoelectron spectroscopic transitions, structure has not been resolved. Many of these states are dissociative. Further information on the dissociation products can be obtained from values of the appearance potentials for various products in photoionization studies on the parent molecule. Such studies are beyond the scope of this review. The tables of published ionization and appearance potentials by Lias and co-workers and the on-line version included in the Chemistry WebBook constitute a valuable source of information on the appearance potentials of photofragments.

Matrix Shifts in Electronic Spectra

The range of tunability of visible and ultraviolet lasers, like that of infrared lasers, is limited. Therefore, a preliminary survey using conventional gas-phase and/or matrix-isolation spectroscopic studies is often desirable. A comparison of the positions of the electronic band origins of diatomic molecules in the gas phase and in rare-gas and nitrogen matrices has been published. As in the determination of ground-state vibrational energy levels, neon is the matrix material of choice, with a sharp maximum at 0.0% in the distribution of matrix deviations for valence transitions of covalently bonded molecules. In argon-matrix observations, most such band origins are shifted by less than 2% from the gas-phase values. At the somewhat higher temperatures often used for electronic spectral observations in matrices of the heavier rare gases or of nitrogen, relatively broad phonon bands become prominent. The blue shift of the phonon maximum from the zero-phonon line in absorption measurements, and the red shift in emission measurements, typically amount to approximately 1 to 1.5%. Rydberg transitions of molecules in matrices often are greatly broadened and experience much larger shifts. Further details of the behaviour of electronic transitions of matrix-isolated molecules have previously been discussed.

Guide to the Compilation

Considerable effort has been expended to provide a critical evaluation of the data. However, for many species the available data are meager. The identities of some species have been proposed on the basis of chemical evidence. While such evidence may be quite compelling, it is not definitive. Many examples could be cited in which a spectrum was later reassigned to characteristic impurities in the sample. Where chemical evidence has provided a reasonable basis for the assignment of vibrational or electronic bands to a transient molecule, data have been included in this compilation, in the hope that further testing of the assignment will be facilitated.

While every effort has been made to make these tables as complete as possible, for various reasons omissions do occur. There remains some selectivity in the coverage of electronic spectral data for larger molecules. Where low-resolution photoelectron spectral data have been superseded by spectroscopic observations with appreciably higher resolution and greater precision, often the photoelectron data have not been cited. Candidate molecules or energy levels may also have been inadvertently omitted. Suggestions of additions or needed revisions to the data to be included in subsequent extensions of this database are welcome, as are inquiries regarding new data added after the cutoff date for this compilation.

Molecular Formulas

Molecular formulas are used in this compilation. In order to permit a compact index, an attempt has been made to provide as much structural information as possible in a minimal amount of space. This restriction is especially severe for larger molecules. The following formula abbreviations have been used:

br		***bridged***
cyc		cyclic. If parentheses follow, only the atoms enclosed in them are included in the ring.
c		*cis*
t		*trans*

Isotopic Species

Where heavy isotopic peaks are resolved, data are given for the most abundant isotopic species (e.g., ^{7}Li, ^{11}B, ^{35}Cl, ^{79}Br). Data are included for both the normal and the fully deuterium-substituted molecule, except that data are also given for the partially deuterium-substituted species of H_3^+ and H_3.

Units

Except where otherwise indicated, the units of all quantities in these tables are cm^{-1}.

Error Estimates

Error estimates are those of the authors of the original literature.

When the uncertainty is not explicitly indicated, the value is given to the estimated number of significant figures. Where vibrational frequencies have been determined with a precision greater than two decimal places, the tabulated values have been rounded off.

Excited Electronic States

Symmetry

The heading for each electronic state gives its symmetry, the point group to which the molecule belongs in that electronic state, and, where available, references to the determination of a quantitative molecular structure. For C_{2v} molecules, there is potential ambiguity in the definition of the molecular symmetry axes. The convention in which the x axis is chosen perpendicular to the plane of the molecule, recommended by the Joint Commission for Spectroscopy of IAU and IUPAP, has been adopted. Often this has required the interchange of published assignments of energy levels with B_1 and B_2 symmetry.

Assignments Based on Photoelectron Spectroscopy

Most authors of papers on photoelectron spectroscopy have proposed assignments for the various photoelectron bands, using arguments based on molecular orbital theory and often on semiempirical or *ab initio* calculations. These assignments have been included in the present tables. Where several conflicting assignments have been given in the literature, an attempt has been made to choose the most satisfactory one.

Generally, the assignments of photoelectron spectra have been made with the presumption that the point group to which the molecular cation belongs is the same in all of its excited states. Structural data for these excited states are extremely rare. Therefore, the molecular point group which has been adopted in the analysis of the photoelectron spectrum is given in these tables. In practice, it is likely that there is some variation in excited-state molecular symmetries. Thus, a bent molecular ion may become linear in some of its excited states. For highly symmetric species, Jahn-Teller distortion may reduce the molecular symmetry.

Energy of the Electronic Transition

The energy of the electronic transition follows the state designation and symmetry information. Where possible, T_0, the energy separation between the electronic energy level of interest and the ground electronic, vibrational, and rotational states of the molecule, is given. However, where only low resolution data or photoelectron data are available, often only band maxima have been given in the literature.

With photoelectron data, T_0 is derived by subtracting the value of the first ionization potential from that of the higher ionization potential which corresponds to the state of interest. When data for the first adiabatic ionization potential are available, the subscript "x" is used to indicate that the first adiabatic ionization potential is known but that the higher ionization potential is measured to the peak maximum; the subscript "xx" implies that the energy difference between the higher and the first absorption maximum was used. If the first photoelectron transition has a gradual onset, a better value of the first ionization potential may have been obtained from photoionization data or from the extrapolation of Rydberg series in the spectrum of the parent molecule.

Supplementary sources of data for the first ionization potential are cited in the tables. However, if the difference between the first adiabatic ionization potential obtained in the photoelectron spectrum and that obtained in other measurements amounts to only 10 or 20 meV, the photoelectron spectroscopic value is used, because of the advantage of a consistent set of measurements.

Where threshold energies differ by one quantum in a vibrational progression, a best value for the ionization potential is chosen which coincides with the most probable position of the vibrationally unexcited transition. Because of inherent uncertainties in the determination of higher ionization potentials in many photoelectron spectral measurements, photoelectron peaks above about 18 eV are often omitted.

As in the tables of Herzberg, T_0 values are given to the centre of multiplet structure. For doublet states, the two components differ by ±A (the spin-orbit splitting constant), and the energy difference is measured from the average of the two bands, whereas for triplet states the three components fall at 0, ±A with respect to the position from which the band energy is measured. This convention is also followed here unless specific states are given. However, in matrix isolation absorption and laser excitation studies only the lowest component is accessible. Except for transitions with relatively small values of A, this is also likely to be true for studies using cooled molecular beams. Often these latter studies give T_0 values for the lowest energy component with a precision better than that to which A is known.

Wavelength Range of the Electronic Transition

The wavelength range (nm) in which various electronic transitions has been observed is also tabulated. This range is a composite of the values typical of absorption and emission observations. Laser-excited fluorescence studies often include both excitation and resolved emission measurements. Since the position of the band origin is given, ambiguity should not arise.

Ground- and Excited-State Vibrations

Vibration Numbering Conventions

The format of the vibrational tables is similar to that used in the earlier compilations. The vibrational numbering convention is that used by Herzberg. Within a given symmetry species, vibrations are numbered starting with the highest frequency. The same convention is followed for deuterated species. Therefore, a given type of vibration may be numbered differently for the deuterated than for the unsubstituted molecule. For triatomic molecules, the bending vibration is always designated as ν_2. For aromatic molecules, an alternate vibrational numbering scheme developed by Wilson has often been used in the literature. Where both the Herzberg and the Wilson numbering schemes have been used for the published data, the Herzberg numbering is adopted, and the Wilson numbering is sometimes shown in parentheses. For a few species, only the Wilson numbering has been used. To avoid confusion, this is retained in the present tables, and the use of the Wilson numbering is specifically indicated. Where possible, the values of $\Delta G(1/2)$, the separation between the $v = 0$ and $v = 1$ levels for the vibration of interest, have been used.

Renner-Teller Interaction

If a bending fundamental is split by Renner-Teller interaction, the position of the unperturbed fundamental is given. Where specific components of such a split fundamental have been studied, they may also be listed, with the transition designated in an explanatory note.

Inversion Splitting

A few of the species in these tables possess out-of-plane vibrations that have resolved inversion splitting structure. For these, the specific component for which the vibrational frequency is reported is designated in an explanatory note.

Intensities

Relative intensities of vibrational bands are dependent on the technique used for the measurement. When possible, the relative intensities of *ground-state infrared absorptions* are included. It is not feasible to give the corresponding relative intensities for other types of observation. Relative intensity abbreviations include:

vw		very weak
w		weak
m		medium
s		strong
vs		very strong
sh		shoulder
br		broad

Abbreviations

Standard Abbreviations for Transition Energies

One or more of five standard abbreviations may be associated with vibrational or electronic transition energies:

H (1/2)(2ν)

I Component of an inversion doublet

L Lower bound

T Tentative assignment or approximate value

U Upper bound

The following abbreviations are used to designate the type of energy separation measured for electronic transitions:

o Energy separation between the v = 0 levels of the excited and the ground electronic states

x Energy separation between the band maximum of the excited electronic state and the v = 0 level of the ground state

xx Energy separation between the band maxima of both the excited state and the ground state of a molecular cation, derived from a photoelectron spectrum

d Photodissociation or photodetachment threshold

a Separation of band origin for an excited state of H_3 or D_3 from the lowest bound state, 2s $^2A_1'$

t Energy separation between the v = 0 levels of the two excited electronic states involved in the transition.

Occasionally, when the band origin is too weak to be observed, specific vibrational quantum numbers are given (e.g., 030 for the A state of D_2O^+).

For ground-state vibrational energy levels of matrix-isolated molecules, the wavenumber column may provide information on species trapped in a site adjacent to the molecule of interest. The method of preparation of a transient molecule may lead to the trapping of, for example, N_2 or HF next to the transient molecule, with consequent shifts in the observed vibrational frequencies.

Molecular anions have sometimes been stabilised by charge transfer interaction between an alkali metal and a suitable precursor molecule. Usually the vibrational frequencies observed for the anion show a small dependence on the nature of the alkali metal present in the system. Where such interactions are known to be significant, the wavenumber, intensity, and standard abbreviations are followed by a space, then by the formula of the perturbing species. (For the charge transfer systems, the perturber is the alkali metal cation. However, the entry is abbreviated to the atomic symbol for the alkali metal present in the sample.)

Type of Measurement

Many sophisticated laser techniques—frequently employing two or more laser beams—have been used for studies of transient molecules. The laser may be used both in the preparation of the transient molecule and in the detection scheme.

For example, ions may be generated by multiphoton ionization and detected by absorption of radiation from a probe laser. Often the developers of such techniques have designated them by complicated acronyms. In these tables, an attempt has been made to avoid relatively lengthy and unfamiliar acronyms by designating only the generic type of detection, using the abbreviations defined below. (Velocity modulation, designated as a separate detection technique in the first of this series of data evaluations, is widely used and is considered to be a measurement tool rather than a type of observation. The type of laser used for the absorption measurement in an infrared detection scheme employing velocity modulation is instead specified in these tables.)

Molecular Electronic Transition

Molecular electronic transitions take place when electrons in a molecule are excited from one energy level to a higher energy level. The energy change associated with this transition provides information on the structure of a molecule and determines many molecular properties such as colour. The relationship between the energy involved in the electronic transition and the frequency of radiation is given by Planck's law.

Organic Molecules and Other Molecules

The electronic transitions in organic compounds and some other compounds can be determined by ultraviolet-visible spectroscopy, provided that transitions in the ultraviolet (UV) or visible range of the electromagnetic spectrum exists for this compound. Electrons occupying a HOMO of a sigma bond can get excited to the LUMO of that bond. This process is denoted as a $\sigma \rightarrow \sigma^*$ transition. Likewise promotion of an electron from a π-bonding orbital to an antibonding π orbital* is denoted as a $\pi \rightarrow \pi^*$ transition. Auxochromes with free electron pairs denoted as n have their own transitions, as do aromatic pi bond transitions. Sections of molecules which can undergo such detectable electron transitions can be referred to as chromophores since such transitions absorb electromagnetic radiation (light), which may be hypothetically perceived as colour somewhere in the electromagnetic spectrum. The following molecular electronic transitions exist:

$\sigma \rightarrow \sigma^*$

$\pi \rightarrow \pi^*$

$n \rightarrow \sigma^*$

$n \rightarrow \pi^*$

aromatic $\pi \rightarrow$ aromatic π^*

In addition to these assignments, electronic transitions also have so-called bands associated with them. The following bands are defined: the R-band from the German *radikalartig* or radical-like, the K-band from the German *Konjugierte* or conjugated, B-band from benzoic and the E-band from ethylenic (system devised by A. Burawoy in 1930). For example, the absorption spectrum for ethane shows a $\sigma \rightarrow \sigma^*$ transition at 135 nm and that of water a $n \rightarrow \sigma^*$ transition at 167 nm with an extinction coefficient of 7,000. Benzene has three aromatic

$\pi \rightarrow \pi^*$ transitions; two E-bands at 180 and 200 nm and one B-band at 255 nm with extinction coefficients respectively 60,000, 8,000 and 215. These absorptions are not narrow bands but are generally broad because the electronic transitions are superimposed on the other molecular energy states.

Solvent Shifts

The electronic transitions of molecules in solution can depend strongly on the type of solvent with additional bathochromic shifts or hypsochromic shifts.

Line Spectra

Spectral lines are associated with atomic electronic transitions and polyatomic gases have their own absorption band system.

Atomic Electron Transition

In physics and chemistry, an atomic electron(ic) transition (could be simply called an atomic transition, a quantum jump, or quantum leap) is a change of an electron from one quantum state to another within an atom (or artificial atom). It appears to be discontinuous (observed for example in fluorescence signal of another transition as burst noise); the electron "jumps" from one energy level to another very quickly (transition time often in a few ns or less). Quantum leaps cause the emission (or absorption) of electromagnetic radiation, including that of light, which occurs in the form of quantized units called photons. Their statistics is Poissonian and the damping of statistic values of time between jumps is exponential (in average). The damping time constant (ranges from nanoseconds to a few seconds) relates to the natural, pressure and field broadening of spectral lines. Although changes of quantum state occur on the submicroscopic level, in popular discourse, the term "quantum leap" refers to a large increase.

Chapter 8

Proton Magnetic Resonance

Proton NMR

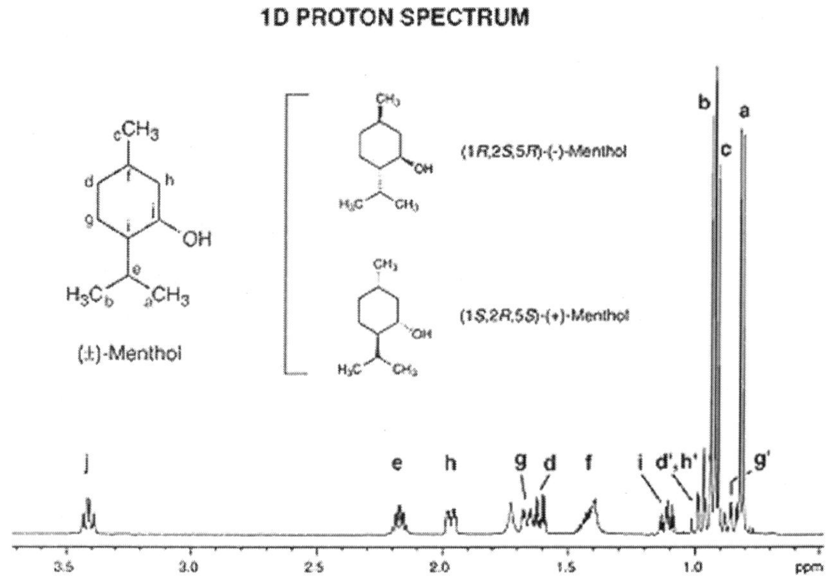

Figure:

Example ^{1}H NMR spectrum (1-dimensional) of a mixture of menthol enantiomers plotted as signal intensity (vertical axis) vs. chemical shift (in ppm on the horizontal axis). Signals from spectrum have been assigned hydrogen atom groups (a through j) from the structure shown at upper left.

Example ^{1}H NMR spectrum (1-dimensional) of a mixture of menthol enantiomers plotted as signal intensity (vertical axis) vs. chemical shift (in ppm on the horizontal axis). Signals from spectrum have been assigned hydrogen atom groups (a through j) from the structure shown at upper left.

Proton NMR (also Hydrogen-1 NMR, or ^{1}H NMR) is the application of nuclear magnetic resonance in NMR spectroscopy with respect to hydrogen-1 nuclei within the molecules of a substance, in order to determine the structure of its molecules. In samples where natural hydrogen (H) is used, practically all of the hydrogen consists of the isotope ^{1}H (hydrogen-1; i.e. having a proton for a nucleus). A full ^{1}H atom is called protium. Simple NMR spectra are recorded in solution, and solvent protons must not be allowed to interfere. Deuterated (deuterium = ^{2}H, often symbolized as D) solvents especially for use in NMR are preferred, e.g. deuterated water, D_2O, deuterated acetone, $(CD_3)_2CO$, deuterated methanol, CD_3OD, deuterated dimethyl sulfoxide, $(CD_3)_2SO$, and deuterated chloroform, $CDCl_3$. However, a solvent without hydrogen, such as carbon tetrachloride, CCl_4 or carbon disulphide, CS_2, may also be used.

Historically, deuterated solvents were supplied with a small amount (typically 0.1 %) of tetramethylsilane (TMS) as an internal standard for calibrating the chemical shifts of each analyte proton. TMS is a tetrahedral molecule, with all protons being chemically equivalent, giving one single signal, used to define a chemical shift = 0 ppm. It is volatile, making sample recovery easy as well. Modern spectrometres are able to reference spectra based on the residual proton in the solvent (e.g. the $CHCl_3$, 0.01 % in 99.99 % $CDCl_3$). Deuterated solvents are now commonly supplied without TMS.

Deuterated solvents permit the use of deuterium frequency-field lock (also known as deuterium lock or field lock) to offset the effect of the natural drift of the NMR's magnetic field B_0. In order to provide deuterium lock, the NMR constantly monitors the deuterium signal resonance frequency from the solvent and makes changes to the to keep the resonance frequency constant. Additionally, the deuterium signal may be used to accurately define 0 ppm as the resonant frequency of the lock solvent and the difference between the lock solvent and 0 ppm (TMS) are well known.

Proton NMR spectra of most organic compounds are characterised by chemical shifts in the range +14 to -4 ppm and by spin-spin coupling between protons. The integration curve for each proton reflects the abundance of the individual protons. Simple molecules have simple spectra. The spectrum of ethyl chloride consists of a triplet at 1.5 ppm and a quartet at 3.5 ppm in a 3:2 ratio. The spectrum of benzene consists of a single peak at 7.2 ppm due to the diamagnetic ring current. Together with Carbon-13 NMR, proton NMR is a powerful tool for molecular structure characterisation.

Chemical Shifts

Chemical shift values, symbolized by δ, are not precise, but typical - they are to be therefore regarded mainly as orientational. Deviations are in ±0.2 ppm range, sometimes more. The exact value of chemical shift depends on molecular structure and the solvent in which the spectrum is being recorded. Hydrogen nuclei are sensitive to the hybridization of the atom to which the hydrogen atom is attached and to electronic effects. Nuclei tend to be deshielded by groups which withdraw electron density. Deshielded nuclei resonate at higher δ values, whereas shielded nuclei resonate at lower δ values.

Examples of electron withdrawing substituents are -OH, -OCOR, -OR, $-NO_2$ and halogens. These cause a downfield shift of approximately 2-4 ppm for H atoms on C_α and of less than 1-2 ppm for H atoms on C_β. C_α is an aliphatic C atom directly bonded to the substituent in question, and C_β is an aliphatic C atom bonded to C_α.

Carbonyl groups, olefinic fragments and aromatic rings contribute sp^2 hybridized carbon atoms to an aliphatic chain. This causes a downfield shift of 1-2 ppm at C_α. Note that labile protons (-OH, $-NH_2$, -SH) have no characteristic chemical shift. However such resonances can be identified by the disappearance of a peak when reacted with D_2O, as deuterium will replace a protium atom. This method is called a D_2O shake. Acidic protons may also be suppressed when a solvent containing acidic deuterium ions (e.g. methanol-d_4) is used.

Functional Group	*CH_3*	*CH_2*	*CH*
CH_2R	0.8	1.3	1.6
C=C	1.6	2.0	2.6
Ca"C	1.7	2.2	2.8
C_6H_5	2.3	2.6	2.9
F	4.3	4.4	4.8
Cl	3.0	3.4	4.0
Br	2.7	3.4	4.1
I	2.2	3.2	4.2
OH	3.3	3.5	3.8
OR	3.3	3.4	3.7
OC_6H_5	3.8	4.0	4.3

OCOR	3.6	4.1	5.0
$OCOC_6H_5$	3.9	4.2	5.1
$OCOCF_3$	4.0	4.4	/
CHO	2.2	2.4	2.5
COR	2.1	2.2	2.6
COOH	2.1	2.3	2.6
COOR	2.0	2.3	2.5
$CONR_2$	2.0	2.1	2.4
CN	2.1	2.5	3.0
NH_2	2.5	2.7	3.0
NR_2	2.2	2.4	2.8
NRC_6H_5	2.6	3.0	3.6
NR_3^+	3.0	3.1	3.6
NHCOR	2.9	3.3	3.7
NO_2	4.1	4.2	4.4
SR	2.1	2.5	3.1
SOR	2.6	3.1	/
=O (aliphatic aldehyde)	/	/	9.5
=O (aromatic aldehyde)	/	/	10
M-H (metal hydride)	/	/	

Spin-Spin Couplings

The chemical shift is not the only indicator used to assign a molecule. Because nuclei themselves possess a small magnetic field, they influence each other, changing the energy and hence frequency of nearby nuclei as they resonate—this is known as spin-spin coupling. The most important type in basic NMR is *scalar coupling*. This interaction between two nuclei occurs through chemical bonds, and can typically be seen up to three bonds away.

The effect of scalar coupling can be understood by examination of a proton which has a signal at 1ppm. This proton is in a hypothetical molecule where three bonds away exists another proton (in a CH-CH group for instance), the neighbouring group (a magnetic field) causes the signal at 1 ppm to split into two, with one peak being a few hertz higher than 1 ppm and the other peak being the same number of hertz lower than 1 ppm. These peaks each have half the area of the former singlet peak. The magnitude of this splitting (difference in frequency between peaks) is known as the coupling constant. A typical coupling constant value would be 7 Hz.

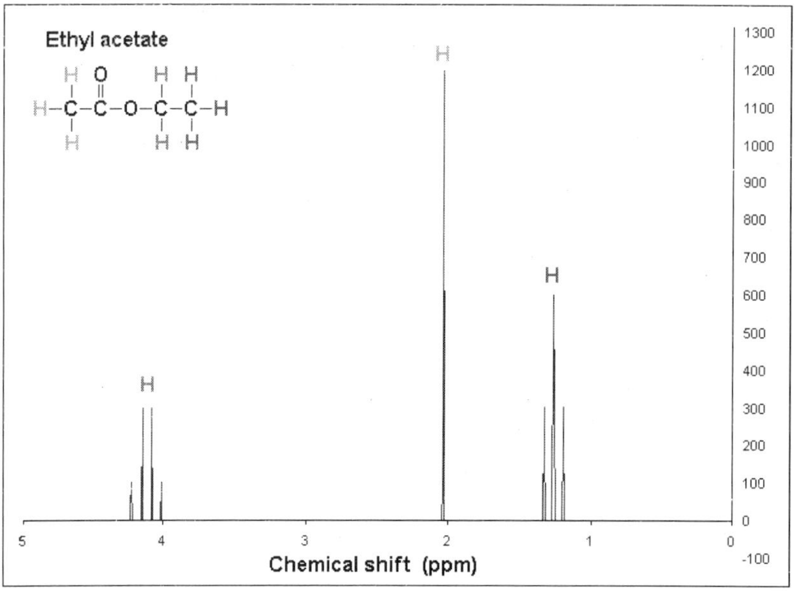

Figure: Example ^{1}H NMR spectrum (1-dimensional) of ethyl acetate plotted as signal intensity vs. chemical shift. There are three different types of H atoms in ethyl acetate regarding NMR. The hydrogens (H) on the CH_3COO- (acetate) group are not coupling with the other H atoms and appear as a singlet, but the $-CH_2-$ and $-CH_3$ hydrogens of the ethyl group ($-CH_2CH_3$) are coupling with each other, resulting in a quartet and triplet respectively.

The coupling constant is independent of magnetic field strength because it is caused by the magnetic field of another nucleus, not the spectrometre magnet. Therefore it is quoted in hertz (frequency) and not ppm (chemical shift). In another molecule a proton resonates at 2.5 ppm and that proton would also be split into two by the proton at 1 ppm. Because the magnitude of interaction is the same the splitting would have the same coupling constant 7 Hz apart. The spectrum would have two signals, each being a doublet. Each doublet will have the same area because both doublets are produced by one proton each.

The two doublets at 1 ppm and 2.5 ppm from the fictional molecule CH-CH are now changed into CH_2-CH:

- The total area of the 1 ppm CH_2 peak will be twice that of the 2.5 ppm CH peak.
- The CH_2 peak will be split into a doublet by the CH peak—with one peak at 1 ppm + 3.5 Hz and one at 1 ppm - 3.5 Hz (total splitting or coupling constant is 7 Hz).

In consequence the CH peak at 2.5 ppm will be split *twice* by each proton from the CH_2. The first proton will split the peak into two equal intensities and will go from one peak at 2.5 ppm to two peaks, one at 2.5 ppm + 3.5 Hz and the other at 2.5 ppm - 3.5 Hz—each having equal intensities. However these will be split again by the second proton. The frequencies will change accordingly:

- The 2.5 ppm + 3.5 Hz signal will be split into 2.5 ppm + 7 Hz and 2.5 ppm
- The 2.5 ppm - 3.5 Hz signal will be split into 2.5 ppm and 2.5 ppm - 7 Hz

The net result is not a signal consisting of 4 peaks but three: one signal at 7 Hz above 2.5 ppm, two signals occur at 2.5 ppm, and a final one at 7 Hz below 2.5 ppm. The ratio of height between them is 1:2:1. This is known as a triplet and is an indicator that the proton is three-bonds from a CH_2 group. This can be extended to any CH_n group. When the CH_2-CH group is changed to CH_3-CH_2, keeping the chemical shift and coupling constants identical, the following changes are observed:

- The relative areas between the CH_3 and CH_2 subunits will be 3:2.
- The CH_3 is coupled to two protons into a 1:2:1 triplet around 1 ppm.
- The CH_2 is coupled to *three* protons.

Something split by three identical protons takes a shape known as a quartet, each peak having relative intensities of 1:3:3:1.

A peak is split by *n* identical protons into components whose sizes are in the ratio of the *n*th row of Pascal's triangle:

n		
1	singlet	1
2	doublet	1 1
3	triplet	1 2 1
4	quartet	1 3 3 1
5	pentet	1 4 6 4 1
6	sextet	1 5 10 10 5 1
7	septet	1 6 15 20 15 6 1
8	octet	1 7 21 35 35 21 7 1
9	nonet	1 8 28 56 70 56 28 8 1

Because the *n*th row has *n*+1 components, this type of splitting is said to follow the "*n*+1 rule": a proton with *n* neighbours appears as a cluster of *n*+1 peaks.

With 2-methylpropane, $(CH_3)_3CH$, as another example: the CH proton is attached to three identical methyl groups. The C-H signal in the spectrum would be split into ten peaks according to the (n + 1) rule of multiplicity. Below are NMR signals corresponding to several simple multiplets of this type. Note that the outer lines of the nonet (which are only 1/8 as high as those of the second peak) can barely be seen, giving a superficial resemblance to a septet.

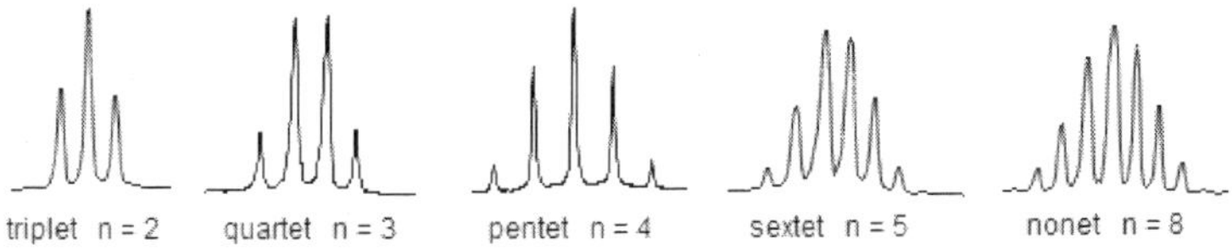

Figure:

When a proton is coupled to two different protons, then the coupling constants are likely to be different, and instead of a triplet, a doublet of doublets will be seen. Similarly, if a proton is coupled to two other protons of one type, and a third of another type with a different, smaller coupling constant, then a triplet of doublets is seen. In the example below, the triplet coupling constant is larger than the doublet one. By convention the pattern created by the largest coupling constant is indicated first and the splitting patterns of smaller constants are named in turn. In the case below it would be erroneous to refer to the quartet of triplets as a triplet of quartets. The analysis of such multiplets (which can get very much more complicated than the ones shown here) provides important clues to the structure of the molecule being studied.

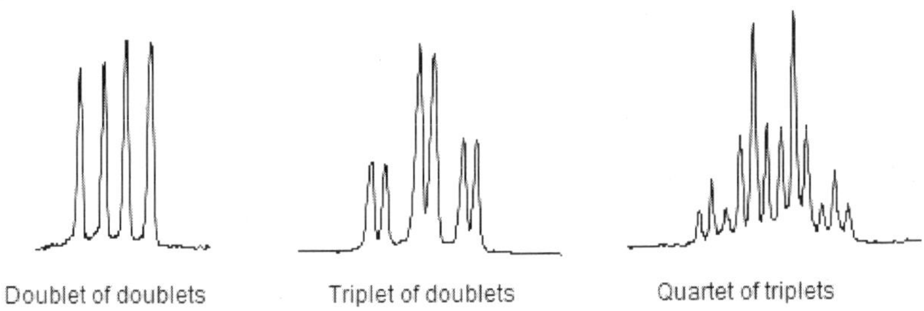

Figure:

The simple rules for the spin-spin splitting of NMR signals described above apply only if the chemical shifts of the coupling partners are substantially larger than the coupling constant between them. Otherwise there may be more peaks, and the intensities of the individual peaks will be distorted (second-order effects).

Carbon Satellites and Spinning Sidebands

Occasionally, small peaks can be seen shouldering the main ^{1}H NMR peaks. These peaks are not the result of proton-proton coupling, but result from the coupling of ^{1}H atoms to an adjoining carbon-13 (^{13}C) atom. These small peaks are known as carbon satellites as they are small and appear around the main ^{1}H peak i.e. satellite (around) to them. Carbon satellites are small because ^{13}C only makes up 1.1% of the atomic carbon content of carbon, the rest of the carbon atoms are predominantly NMR inactive Carbon-12. Carbon satellites always appear as an evenly spaced pair around the main ^{1}H peak.

This is because they are the result of 1.1% of the ^{1}H atoms coupling to an adjoined ^{13}C atom to give a wide doublet (^{13}C has a spin of a half). Note, if the main ^{1}H peak has proton-proton coupling, then each satellite will be a miniature version of the main peak and will also show this ^{1}H coupling, e.g. if the main ^{1}H peak is a doublet, then the carbon satellites will appear as miniature doublets, i.e. one doublet on either side of the main ^{1}H peak. Sometime other peaks can be seen around ^{1}H peaks, these are known as spinning sidebands and are related to the rate of spin of an NMR tube. Carbon satellites and spinning sidebands should not be confused with impurity peaks.

Mass Spectrometry

Mass spectrometry (MS) is an analytical technique that measures the mass-to-charge ratio of charged particles. It is used for determining masses of particles, for determining the elemental composition of a sample or molecule, and for elucidating the chemical structures of molecules, such as peptides and other chemical compounds. MS works by ionizing chemical compounds to generate charged molecules or molecule fragments and measuring their mass-to-charge ratios. In a typical MS procedure:

1. A sample is loaded onto the MS instrument and undergoes vapourisation
2. The components of the sample are ionized by one of a variety of methods (e.g., by impacting them with an electron beam), which results in the formation of charged particles (ions)

3. The ions are separated according to their mass-to-charge ratio in an analyser by electromagnetic fields
4. The ions are detected, usually by a quantitative method
5. The ion signal is processed into mass spectra

MS instruments consist of three modules:

- An *ion source*, which can convert gas phase sample molecules into ions (or, in the case of electrospray ionization, move ions that exist in solution into the gas phase)
- A *mass analyser*, which sorts the ions by their masses by applying electromagnetic fields
- A *detector*, which measures the value of an indicator quantity and thus provides data for calculating the abundances of each ion present

The technique has both qualitative and quantitative uses. These include identifying unknown compounds, determining the isotopic composition of elements in a molecule, and determining the structure of a compound by observing its fragmentation. Other uses include quantifying the amount of a compound in a sample or studying the fundamentals of gas phase ion chemistry (the chemistry of ions and neutrals in a vacuum). MS is now in very common use in analytical laboratories that study physical, chemical, or biological properties of a great variety of compounds.

Etymology

The word *spectrograph* had become part of the international scientific vocabulary by 1884. The linguistic roots are a combination and removal of bound morphemes and free morphemes which relate to the terms *spectr-um* and *phot-ograph-ic plate*. Early *spectrometry* devices that measured the mass-to-charge ratio of ions were called *mass spectrographs* which consisted of instruments that recorded a spectrum of mass values on a photographic plate. A *mass spectroscope* is similar to a *mass spectrograph* except that the beam of ions is directed onto a phosphor screen.

A mass spectroscope configuration was used in early instruments when it was desired that the effects of adjustments be quickly observed. Once the instrument was properly adjusted, a photographic plate was inserted and exposed. The term mass spectroscope continued to be used even though the direct illumination of a phosphor screen was replaced by indirect measurements with an oscilloscope. The use of the term *mass spectroscopy* is now discouraged due to the possibility of confusion with light spectroscopy. Mass spectrometry is often abbreviated as *mass-spec* or simply as *MS.*

In 1886, Eugen Goldstein observed rays in gas discharges under low pressure that travelled away from the anode and through channels in a perforated cathode, opposite to the direction of negatively charged cathode rays (which travel from cathode to anode). Goldstein called these positively charged anode rays "Kanalstrahlen"; the standard translation of this term into English is "canal rays". Wilhelm Wien found that strong electric or magnetic fields deflected the canal rays and, in 1899, constructed a device with parallel electric and magnetic fields that separated the positive rays according to their charge-to-mass ratio (Q/m). Wien found that the charge-to-mass ratio depended on the nature of the gas in the discharge tube. English scientist J.J. Thomson later improved on the work of Wien by reducing the pressure to create the mass spectrograph.

The first application of mass spectrometry to the analysis of amino acids and peptides was reported in 1958. Carl-Ove Andersson highlighted the main fragment ions observed in the ionization of methyl esters.

Some of the modern techniques of mass spectrometry were devised by Arthur Jeffrey Dempster and F.W. Aston in 1918 and 1919 respectively. In 1989, half of the Nobel Prize in Physics was awarded to Hans Dehmelt and Wolfgang Paul for the development of the ion trap technique in the 1950s and 1960s. In 2002, the Nobel Prize in Chemistry was awarded to John Bennett Fenn for the development of electrospray ionization (ESI) and Koichi Tanaka for the development of soft laser desorption (SLD) and their application to the ionization of biological macromolecules, especially proteins.

Simplified Example

The following example describes the operation of a spectrometre mass analyser, which is of the sector type. (Other analyser types are treated below.) Consider a sample of sodium chloride. In the ion source, the sample is vapourised (turned into gas) and ionized (transformed into electrically charged particles) into sodium (Na^+) and chloride (Cl^-) ions. Sodium atoms and ions are monoisotopic, with a mass of about 23 amu. Chloride atoms and ions come in two isotopes with masses of approximately 35 amu (at a natural abundance of about 75 percent) and approximately 37 amu (at a natural abundance of about 25 percent). The analyser part of the spectrometre contains electric and magnetic fields, which exert forces on ions travelling through these fields.

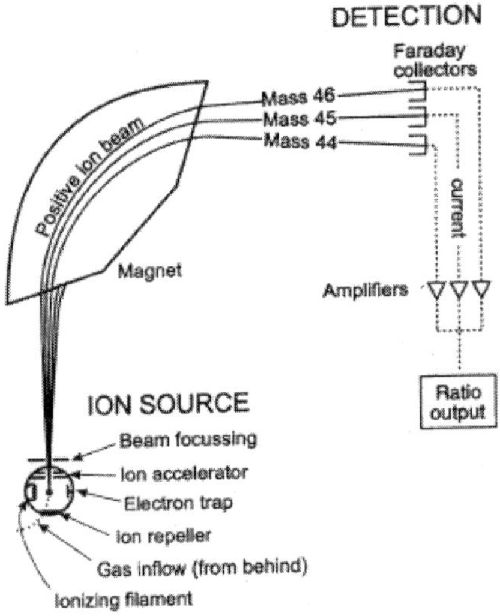

Figure: Schematics of a simple mass spectrometre with sector type mass analyser. This one is for the measurement of carbon dioxide isotope ratios (IRMS) as in the carbon-13 urea breath test

The speed of a charged particle may be increased or decreased while passing through the electric field, and its direction may be altered by the magnetic field. The magnitude of the deflection of the moving ion's trajectory depends on its mass-to-charge ratio. Lighter ions get deflected by the magnetic force more than heavier ions (based on Newton's second law of motion, F = ma). The streams of sorted ions pass from the analyser to the detector, which records the relative abundance of each ion type. This information is used to determine the chemical element composition of the original sample (i.e. that both sodium and chlorine are present in the sample) and the isotopic composition of its constituents (the ratio of ^{35}Cl to ^{37}Cl).

Creating Ions

The ion source is the part of the mass spectrometre that ionizes the material under analysis (the analyte). The ions are then transported by magnetic or electric fields to the mass analyser. Techniques for ionization have been key to determining what types of samples can be analysed by mass spectrometry. Electron ionization and chemical ionization are used for gases and vapours. In chemical ionization sources, the analyte is ionized by chemical ion-molecule reactions during collisions in the source.

Two techniques often used with liquid and solid biological samples include electrospray ionization (invented by John Fenn) and matrix-assisted laser desorption/ionization (MALDI, initially developed as a similar technique "Soft Laser Desorption (SLD)" by K. Tanaka for which a Nobel Prize was awarded and as MALDI by M. Karas and F. Hillenkamp).

Inductively Coupled Plasma

Inductively coupled plasma (ICP) sources are used primarily for cation analysis of a wide array of sample types. In this type of Ion Source Technology, a 'flame' of plasma that is electrically neutral overall, but that has had a substantial fraction of its atoms ionized by high temperature, is used to atomize introduced sample molecules and to further strip the outer electrons from those atoms. The plasma is usually generated from argon gas, since the first ionization energy of argon atoms is higher than the first of any other elements except He, O, F and Ne, but lower than the second ionization energy of all except the most electropositive metals. The heating is achieved by a radio-frequency current passed through a coil surrounding the plasma.

Other Ionization Techniques

Others include glow discharge, field desorption (FD), fast atom bombardment (FAB), thermospray, desorption/ionization on silicon (DIOS), Direct Analysis in Real Time (DART), atmospheric pressure chemical ionization (APCI), secondary ion mass spectrometry (SIMS), spark ionization and thermal ionization (TIMS). Ion attachment ionization is an ionization technique that allows for fragmentation free analysis.

Mass Selection

Mass analysers separate the ions according to their mass-to-charge ratio. The following two laws govern the dynamics of charged particles in electric and magnetic fields in vacuum:

$\mathbf{F} = Q(\mathbf{E} + \mathbf{v} \times \mathbf{B})$ (Lorentz force law);

$\mathbf{F} = m\mathbf{a}$ Newton's second law of motion in non-relativistic case, i.e. valid only at ion velocity much lower than the speed of light).

Here F is the force applied to the ion, m is the mass of the ion, a is the acceleration, Q is the ion charge, E is the electric field, and $v \times B$ is the vector cross product of the ion velocity and the magnetic field.

Equating the above expressions for the force applied to the ion yields:

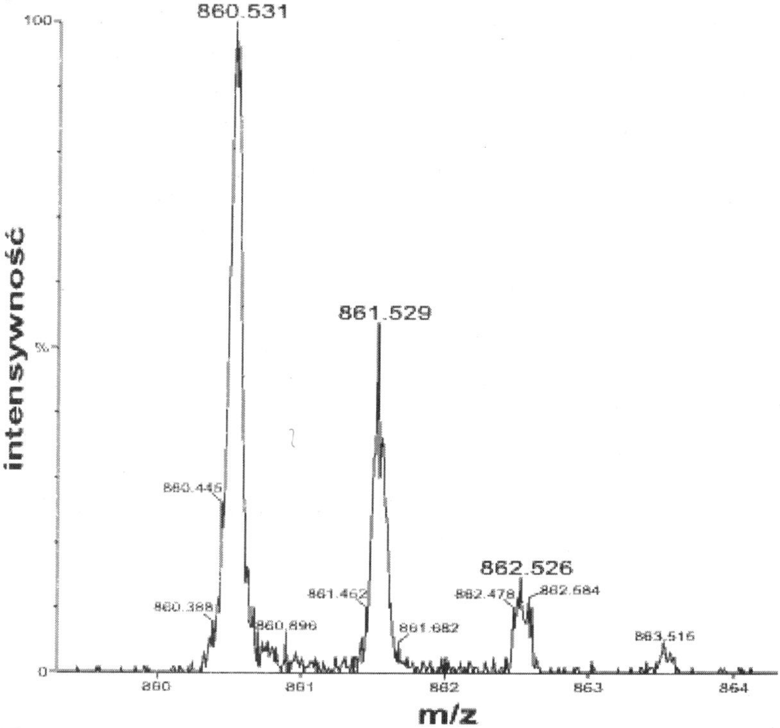

Figure:

This differential equation is the classic equation of motion for charged particles. Together with the particle's initial conditions, it completely determines the particle's motion in space and time in terms of m/Q. Thus mass spectrometres could be thought of as "mass-to-charge spectrometres". When presenting data, it is common to use the (officially) dimensionless m/z, where z is the number of elementary charges (e) on the ion (z=Q/e). This quantity, although it is informally called the mass-to-charge ratio, more accurately speaking represents the ratio of the mass number and the charge number, z.

There are many types of mass analysers, using either static or dynamic fields, or magnetic or electric fields, but all operate according to the above differential equation. Each analyser type has its strengths and weaknesses. Many mass spectrometres use two or more mass analysers for tandem mass spectrometry (MS/MS). In addition to the more common mass analysers listed below, there are others designed for special situations.

There are several important analyser characteristics. The mass resolving power is the measure of the ability to distinguish two peaks of slightly different *m/z*. The mass accuracy is the ratio of the *m/z* measurement error to the true m/z. Mass accuracy is usually measured in ppm or milli mass units. The mass range is the range of *m/z* amenable to analysis by a given analyser. The linear dynamic range is the range over which ion signal is linear with analyte concentration. Speed refers to the time frame of the experiment and ultimately is used to determine the number of spectra per unit time that can be generated.

Sector Instruments

A sector field mass analyser uses an electric and/or magnetic field to affect the path and/or velocity of the charged particles in some way. As shown above, sector instruments bend the trajectories of the ions as they pass through the mass analyser, according to their mass-to-charge ratios, deflecting the more charged and faster-moving, lighter ions more. The analyser can be used to select a narrow range of *m/z* or to scan through a range of *m/z* to catalog the ions present.

Time-of-Flight

The time-of-flight (TOF) analyser uses an electric field to accelerate the ions through the same potential, and then measures the time they take to reach the detector. If the particles all have the same charge, the kinetic energies will be identical, and their velocities will depend only on their masses. Lighter ions will reach the detector first.

Quadrupole Mass Filter

Quadrupole mass analysers use oscillating electrical fields to selectively stabilise or destabilise the paths of ions passing through a radio frequency (RF) quadrupole field created between 4 parallel rods. Only the ions in a certain range of mass/charge ratio are passed through the system at any time, but changes to the potentials on the rods allow a wide range of m/z values to be swept rapidly, either continuously or in a succession of discrete hops. A quadrupole mass analyser acts as a mass-selective filter and is closely related to the quadrupole ion trap, particularly the linear quadrupole ion trap except that it is designed to pass the untrapped ions rather than collect the trapped ones, and is for that reason referred to as a transmission quadrupole.

A common variation of the transmission quadrupole is the triple quadrupole mass spectrometre. The "triple quad" has three consecutive quadrupole stages, the first acting as a mass filter to transmit a particular incoming ion to the second quadrupole, a collision chamber, wherein that ion can be broken into fragments. The third quadrupole also acts as a mass filter, to transmit a particular fragment ion to the detector. If a quadrupole is made to rapidly and repetitively cycle through a range of mass filter settings, full spectra can be reported. Likewise, a triple quad can be made to perform various scan types characteristic of tandem mass spectrometry.

Ion Traps

Three-Dimensional Quadrupole Ion Trap

The quadrupole ion trap works on the same physical principles as the quadrupole mass analyser, but the ions are trapped and sequentially ejected. Ions are trapped in a mainly quadrupole RF field, in a space defined by a ring electrode (usually connected to the main RF potential) between two endcap electrodes (typically connected to DC or auxiliary AC potentials). The sample is ionized either internally (e.g. with an electron or laser beam), or externally, in which case the ions are often introduced through an aperture in an endcap electrode.

There are many mass/charge separation and isolation methods but the most commonly used is the mass instability mode in which the RF potential is ramped so that the orbit of ions with a mass $a > b$ are stable while ions with mass b become unstable and are ejected on the z-axis onto a detector. There are also non-destructive analysis methods.

Ions may also be ejected by the resonance excitation method, whereby a supplemental oscillatory excitation voltage is applied to the endcap electrodes, and the trapping voltage amplitude and/or excitation voltage frequency is varied to bring ions into a resonance condition in order of their mass/charge ratio. The cylindrical ion trap mass spectrometre is a derivative of the quadrupole ion trap mass spectrometre.

Linear Quadrupole Ion Trap

A linear quadrupole ion trap is similar to a quadrupole ion trap, but it traps ions in a two dimensional quadrupole field, instead of a three-dimensional quadrupole field as in a 3D quadrupole ion trap. Thermo Fisher's LTQ ("linear trap quadrupole") is an example of the linear ion trap.

A toroidal ion trap can be visualized as a linear quadrupole curved around and connected at the ends or as a cross section of a 3D ion trap rotated on edge to form the toroid, donut shaped trap. The trap can store large volumes of ions by distributing them throughout the ring-like trap structure. This toroidal shaped trap is a configuration that allows the increased miniaturization of an ion trap mass analyser. Additionally all ions are stored in the same trapping field and ejected together simplifying detection that can be complicated with array configurations due to variations in detector alignment and machining of the arrays.

Orbitrap

These are similar to Fourier transform ion cyclotron resonance mass spectrometres. Ions are electrostatically trapped in an orbit around a central, spindle shaped electrode. The electrode confines the ions so that they both orbit around the central electrode and oscillate back and forth along the central electrode's long axis. This oscillation generates an image current in the detector plates which is recorded by the instrument. The frequencies of these image currents depend on the mass to charge ratios of the ions. Mass spectra are obtained by Fourier transformation of the recorded image currents. Orbitraps have a high mass accuracy, high sensitivity and a good dynamic range.

Fourier Transform Ion Cyclotron Resonance

Fourier transform mass spectrometry (FTMS), or more precisely Fourier transform ion cyclotron resonance MS, measures mass by detecting the image current produced by ions cyclotroning in the presence of a magnetic field. Instead of measuring the deflection of ions with a detector such as an electron multiplier, the ions are injected into a Penning trap (a static electric/magnetic ion trap) where they effectively form part of a circuit. Detectors at fixed positions in space measure the electrical signal of ions which pass near them over time, producing a periodic signal. Since the frequency of an ion's cycling is determined by its mass to charge ratio, this can be deconvoluted by performing a Fourier transform on the signal. FTMS has the advantage of high sensitivity (since each ion is "counted" more than once) and much higher resolution and thus precision.

Ion cyclotron resonance (ICR) is an older mass analysis technique similar to FTMS except that ions are detected with a traditional detector. Ions trapped in a Penning trap are excited by an RF electric field until they impact the wall of the trap, where the detector is located. Ions of different mass are resolved according to impact time.

Detectors

The final element of the mass spectrometre is the detector. The detector records either the charge induced or the current produced when an ion passes by or hits a surface. In a scanning instrument, the signal produced in the detector during the course of the scan versus where the instrument is in the scan (at what m/Q) will produce a mass spectrum, a record of ions as a function of m/Q.

Typically, some type of electron multiplier is used, though other detectors including Faraday cups and ion-to-photon detectors are also used. Because the number of ions leaving the mass analyser at a particular instant is typically quite small, considerable amplification is often necessary to get a signal. Microchannel plate detectors are commonly used in modern commercial instruments. In FTMS and Orbitraps, the detector consists of a pair of metal surfaces within the mass analyser/ion trap region which the ions only pass near as they oscillate. No DC current is produced, only a weak AC image current is produced in a circuit between the electrodes. Other inductive detectors have also been used.

Tandem Mass Spectrometry

A tandem mass spectrometre is one capable of multiple rounds of mass spectrometry, usually separated by some form of molecule fragmentation. For example, one mass analyser can isolate one peptide from many entering a mass spectrometre. A second mass analyser then stabilises the peptide ions while they collide with a gas, causing them to fragment by collision-induced dissociation (CID). A third mass analyser then sorts the fragments produced from the peptides. Tandem MS can also be done in a single mass analyser over time, as in a quadrupole ion trap. There are various methods for fragmenting molecules for tandem MS, including collision-induced dissociation (CID), electron capture dissociation (ECD), electron transfer dissociation (ETD), infrared multiphoton dissociation (IRMPD), blackbody infrared radiative dissociation (BIRD), electron-detachment dissociation (EDD) and surface-induced dissociation (SID). An important application using tandem mass spectrometry is in protein identification.

Tandem mass spectrometry enables a variety of experimental sequences. Many commercial mass spectrometres are designed to expedite the execution of such routine sequences as selected reaction monitoring (SRM) and precursor ion scanning. In SRM, the first analyser allows only a single mass through and the second analyser monitors for multiple user-defined fragment ions. SRM is most often used with scanning instruments where the second mass analysis event is duty cycle limited. These experiments are used to increase specificity of detection of known molecules, notably in pharmacokinetic studies. Precursor ion scanning refers to monitoring for a specific loss from the precursor ion. The first and second mass analysers scan across the spectrum as partitioned by a user-defined m/z value. This experiment is used to detect specific motifs within unknown molecules.

Another type of tandem mass spectrometry used for radiocarbon dating is accelerator mass spectrometry (AMS), which uses very high voltages, usually in the mega-volt range, to accelerate negative ions into a type of tandem mass spectrometre.

Common Mass Spectrometre Configurations and Techniques

When a specific configuration of source, analyser, and detector becomes conventional in practice, often a compound acronym arises to designate it, and the compound acronym may be better known among nonspectrometrists than the component acronyms. The epitome of this is MALDI-TOF, which simply refers to combining a matrix-assisted laser desorption/ionization source with a time-of-flight mass analyser. The MALDI-TOF moniker is more widely recognised by the non-mass spectrometrists than MALDI or TOF individually. Other examples include inductively coupled plasma-mass spectrometry (ICP-MS), accelerator mass spectrometry (AMS), thermal ionization-mass spectrometry (TIMS) and spark source mass spectrometry (SSMS). Sometimes the use of the generic "MS" actually connotes a very specific mass analyser and detection system, as is the case with AMS, which is always sector based. Certain applications of mass spectrometry have developed monikers that although strictly speaking would seem to refer to a broad application, in practice have come instead to connote a specific or a limited number of instrument configurations. An example of this is isotope ratio mass spectrometry (IRMS), which refers in practice to the use of a limited number of sector based mass analysers; this name is used to refer to both the application and the instrument used for the application.

Chromatographic Techniques Combined with Mass Spectrometry

An important enhancement to the mass resolving and mass determining capabilities of mass spectrometry is using it in tandem with chromatographic separation techniques.

Gas Chromatography

A common combination is gas chromatography-mass spectrometry (GC/MS or GC-MS). In this technique, a gas chromatograph is used to separate different compounds. This stream of separated compounds is fed online into the ion source, a metallic filament to which voltage is applied. This filament emits electrons which ionize the compounds. The ions can then further fragment, yielding predictable patterns. Intact ions and fragments pass into the mass spectrometre's analyser and are eventually detected.

Liquid Chromatography

Similar to gas chromatography MS (GC/MS), liquid chromatography mass spectrometry (LC/MS or LC-MS) separates compounds chromatographically before they are introduced to the ion source and mass spectrometre. It differs from GC/MS in that the mobile phase is liquid, usually a mixture of water and organic solvents, instead of gas and the ions fragments cannot yield predictable patterns. Most commonly, an electrospray ionization source is used in LC/MS. There are also some newly developed ionization techniques like laser spray.

Liquid chromatography–mass spectrometry (LC-MS, or alternatively HPLC-MS) is an analytical chemistry technique that combines the physical separation capabilities of liquid chromatography (or HPLC) with the mass analysis capabilities of mass spectrometry. LC-MS is a powerful technique used for many applications which has very high sensitivity and selectivity. Generally its application is oriented towards the general detection and potential identification of chemicals in the presence of other chemicals (in a complex mixture). The limitations of LC-MS in urine analysis drug screening are that it often fails to distinguish between specific metabolites, in particular with hydrocodone and its metabolites. LC-MS urine analysis testing is used to detect specific categories of drugs. However, gas chromatography (GC-MS) should be used when detection of a specific drug and its metabolites is required.

Liquid Chromatography

Scale

A major difference between traditional HPLC and the chromatography used in LC-MS is that in the latter case the scale is usually much smaller, both with respect to the internal diametre of the column and even more so with respect to flow rate since it scales as the square of the diametre. For a long time, 1 mm columns were typical for LC-MS work (as opposed to 4.6 mm for HPLC). More recently 300 μm and even 75 μm capillary columns have become more prevalent. At the low end of these column diametres the flow rates approach 100 nL/min and are generally used with nanospray sources.

Flow Splitting

When standard bore (4.6 mm) columns are used the flow is often split ~10:1. This can be beneficial by allowing the use of other techniques in tandem such as MS and UV. However splitting the flow to UV will decrease the sensitivity of spectrophotometric detectors. The mass spectrometry on the other hand will give improved sensitivity at flow rates of 200 μL/min or less.

Mass Spectrometry

Mass Analyser

There are many different mass analysers that can be used in LC/MS. Single Quadrupole, Triple Quadrupole, Ion Trap, TOF (time of Flight) and Quadrupole-time of flight (Q-TOF).

Interface

Understandably the interface between a liquid phase technique which continuously flows liquid, and a gas phase technique carried out in a vacuum was difficult for a long time. The advent of electrospray ionization changed this. The interface is most often an electrospray ion source or variant such as a nanospray source; however atmospheric pressure chemical ionization interface is also used. Various deposition and drying techniques have also been used such as using moving belts; however the most common of these is off-line MALDI deposition. A new approach still under development called Direct-EI LC-MS interface, couples a nano HPLC system and an electron ionization equipped mass spectrometre.

Pharmacokinetics

LC-MS is very commonly used in pharmacokinetic studies of pharmaceuticals and is thus the most frequently used technique in the field of bioanalysis. These studies give information about how quickly a drug will be cleared from the hepatic blood flow, and organs of the body. MS is used for this due to high sensitivity and exceptional specificity compared to UV (as long as the analyte can be suitably ionised), and short analysis time.

The major advantage MS has is the use of tandem MS-MS. The detector may be programmed to select certain ions to fragment. The process is essentially a selection technique, but is in fact more complex. The measured quantity is the sum of molecule fragments chosen by the operator. As long as there are no interferences or ion suppression, the LC separation can be quite quick. It is common now to have analysis times of 1 minute or less by MS-MS detection, compared to over 10 mins with UV detection.

Proteomics/Metabolomics

LC-MS is also used in the study of proteomics where again components of a complex mixture must be detected and identified in some manner. The bottom-up proteomics LC-MS approach to proteomics generally involves protease digestion and denaturation (usually trypsin as a protease, urea to denature tertiary structure and iodoacetamide to cap cysteine residues) followed by LC-MS with peptide mass fingerprinting or LC-MS/MS (tandem MS) to derive sequence of individual peptides.

LC-MS/MS is most commonly used for proteomic analysis of complex samples where peptide masses may overlap even with a high-resolution mass spectrometre. Samples of complex biological fluids like human serum may be run in a modern LC-MS/MS system and result in over 1000 proteins being identified, provided that the sample was first separated on an SDS-PAGE gel or HPLC-SCX.

Drug Development

LC-MS is frequently used in drug development at many different stages including Peptide Mapping, Glycoprotein Mapping, Natural Products Dereplication, Bioaffinity Screening, In Vivo Drug Screening, Metabolic Stability Screening, Metabolite Identification, Impurity Identification, Degradant Identification, Quantitative Bioanalysis, and Quality Control.

Ion Mobility

Ion mobility spectrometry/mass spectrometry (IMS/MS or IMMS) is a technique where ions are first separated by drift time through some neutral gas under an applied electrical potential gradient before being introduced into a mass spectrometre. Drift time is a measure of the radius relative to the charge of the ion. The duty cycle of IMS (the time over which the experiment takes place) is longer than most mass spectrometric techniques, such that the mass spectrometre can sample along the course of the IMS separation. This produces data about the IMS separation and the mass-to-charge ratio of the ions in a manner similar to LC/MS. The duty cycle of IMS is short relative to liquid chromatography or gas chromatography separations and can thus be coupled to such techniques, producing triple modalities such as LC/IMS/MS.

Data and Analysis

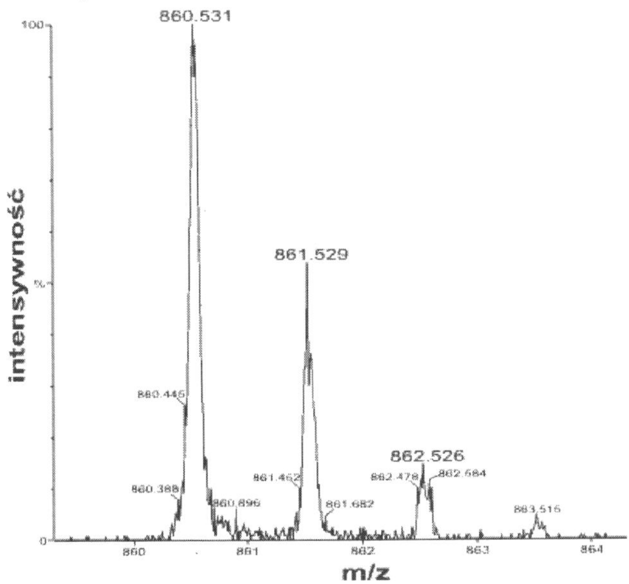

Figure: Mass spectrum of a peptide showing the isotopic distribution

Data Representations

Mass spectrometry produces various types of data. The most common data representation is the mass spectrum. Certain types of mass spectrometry data are best represented as a mass chromatogram. Types of chromatograms include selected ion monitoring (SIM), total ion current (TIC), and selected reaction monitoring chromatogram (SRM), among many others.

Other types of mass spectrometry data are well represented as a three-dimensional contour map. In this form, the mass-to-charge, m/z is on the x-axis, intensity the y-axis, and an additional experimental parameter, such as time, is recorded on the z-axis.

Data Analysis

Basics

Mass spectrometry data analysis is a complicated subject that is very specific to the type of experiment producing the data. There are general subdivisions of data that are fundamental to understanding any data.

Many mass spectrometres work in either *negative ion mode* or *positive ion mode.* It is very important to know whether the observed ions are negatively or positively charged. This is often important in determining the neutral mass but it also indicates something about the nature of the molecules.

Different types of ion source result in different arrays of fragments produced from the original molecules. An electron ionization source produces many fragments and mostly single-charged (1-) radicals (odd number of electrons), whereas an electrospray source usually produces non-radical quasimolecular ions that are frequently multiply charged. Tandem mass spectrometry purposely produces fragment ions post-source and can drastically change the sort of data achieved by an experiment.

By understanding the origin of a sample, certain expectations can be assumed as to the component molecules of the sample and their fragmentations. A sample from a synthesis/manufacturing process will probably contain impurities chemically related to the target component. A relatively crudely prepared biological sample will probably contain a certain amount of salt, which may form adducts with the analyte molecules in certain analyses.

Results can also depend heavily on how the sample was prepared and how it was run/introduced. An important example is the issue of which matrix is used for MALDI spotting, since much of the energetics of the desorption/ionization event is controlled by the matrix rather than the laser power. Sometimes samples are spiked with sodium or another ion-carrying species to produce adducts rather than a protonated species. The greatest source of trouble when non-mass spectrometrists try to conduct mass spectrometry on their own or collaborate with a mass spectrometrist is inadequate definition of the research goal of the experiment. Adequate definition of the experimental goal is a prerequisite for collecting the proper data and successfully interpreting it. Among the determinations that can be achieved with mass spectrometry are molecular mass, molecular structure, and sample purity. Each of these questions requires a different experimental procedure. Simply asking for a "mass spec" will most likely not answer the real question at hand.

Interpretation of Mass Spectra

Since the precise structure or peptide sequence of a molecule is deciphered through the set of fragment masses, the interpretation of mass spectra requires combined use of various techniques. Usually the first strategy for identifying an unknown compound is to compare its experimental mass spectrum against a library of mass spectra. If the search comes up empty, then manual interpretation or software assisted interpretation of mass spectra are performed. Computer simulation of ionization and fragmentation processes occurring in mass spectrometre is the primary tool for assigning structure or peptide sequence to a molecule. An *a priori* structural information is fragmented *in silico* and the resulting pattern is compared with observed spectrum. Such simulation is often supported by a fragmentation library that contains published patterns of known decomposition reactions. Software taking advantage of this idea has been developed for both small molecules and proteins.

Another way of interpreting mass spectra involves spectra with accurate mass. A mass-to-charge ratio value (m/z) with only integer precision can represent an immense number of theoretically possible ion structures. More precise mass figures significantly reduce the number of candidate molecular formulas, albeit each can still represent large number of structurally diverse compounds. A computer algorithm called formula generator calculates all molecular formulas that theoretically fit a given mass with specified tolerance. A recent technique for structure elucidation in mass spectrometry, called precursor ion fingerprinting identifies individual pieces of structural information by conducting a search of the tandem spectra of the molecule under investigation against a library of the product-ion spectra of structurally characterised precursor ions.

Isotope Ratio MS: Isotope Dating and Tracking

Mass spectrometry is also used to determine the isotopic composition of elements within a sample. Differences in mass among isotopes of an element are very small, and the less abundant isotopes of an element are typically very rare, so a very sensitive instrument is required. These instruments, sometimes referred to as isotope ratio mass spectrometres (IR-MS), usually use a single magnet to bend a beam of ionized particles towards a series of Faraday cups which convert particle impacts to electric current. A fast on-line analysis of deuterium content of water can be done using Flowing afterglow mass spectrometry, FA-MS. Probably the most sensitive and accurate mass spectrometre for this purpose is the accelerator mass spectrometre (AMS). Isotope ratios are important markers of a variety of processes. Some isotope ratios are used to determine the age of materials for example as in carbon dating. Labelling with stable isotopes is also used for protein quantification.

Trace Gas Analysis

Several techniques use ions created in a dedicated ion source injected into a flow tube or a drift tube: selected ion flow tube (SIFT-MS), and proton transfer reaction (PTR-MS), are variants of chemical ionization dedicated for trace gas analysis of air, breath or liquid headspace using well defined reaction time allowing calculations of analyte concentrations from the known reaction kinetics without the need for internal standard or calibration.

Atom Probe

An atom probe is an instrument that combines time-of-flight mass spectrometry and field ion microscopy (FIM) to map the location of individual atoms.

Pharmacokinetics

Pharmacokinetics is often studied using mass spectrometry because of the complex nature of the matrix (often blood or urine) and the need for high sensitivity to observe low dose and long time point data. The most common instrumentation used in this application is LC-MS with a triple quadrupole mass spectrometre. Tandem mass spectrometry is usually employed for added specificity. Standard curves and internal standards are used for quantitation of usually a single pharmaceutical in the samples. The samples represent different time points as a pharmaceutical is administered and then metabolized or cleared from the body. Blank or t=0 samples taken before administration are important in determining background and ensuring data integrity with such complex sample matrices. Much attention is paid to the linearity of the standard curve; however it is not uncommon to use curve fitting with more complex functions such as quadratics since the response of most mass spectrometres is less than linear across large concentration ranges. There is currently considerable interest in the use of very high sensitivity mass spectrometry for microdosing studies, which are seen as a promising alternative to animal experimentation.

Protein Characterisation

Mass spectrometry is an important emerging method for the characterisation and sequencing of proteins. The two primary methods for ionization of whole proteins are electrospray ionization (ESI) and matrix-assisted laser desorption/ionization (MALDI). In keeping with the performance and mass range of available mass spectrometres, two approaches are used for characterising proteins. In the first, intact proteins are ionized by either of the two techniques described above, and then introduced to a mass analyser. This approach is referred to as "top-down" strategy of protein analysis. In the second, proteins are enzymatically digested into smaller peptides using proteases such as trypsin or pepsin, either in solution or in gel after electrophoretic separation. Other proteolytic agents are also used. The collection of peptide products are then introduced to the mass analyser. When the characteristic pattern of peptides is used for the identification of the protein the method is called peptide mass fingerprinting (PMF), if the identification is performed using the sequence data determined in tandem MS analysis it is called de novo sequencing. These procedures of protein analysis are also referred to as the "bottom-up" approach.

Protein mass spectrometry refers to the application of mass spectrometry to the study of proteins. Mass spectrometry is an important emerging method for the characterisation of proteins. The two primary methods for ionization of whole proteins are electrospray ionization (ESI) and matrix-assisted laser desorption/ionization (MALDI). In keeping with the performance and mass range of available mass spectrometres, two approaches are used for characterising proteins. In the first, intact proteins are ionized by either of the two techniques described above, and then introduced to a mass analyser. This approach is referred to as "top-down" strategy of protein analysis. In the second, proteins are enzymatically digested into smaller peptides using a protease such as trypsin. Subsequently these peptides are introduced into the mass spectrometre and identified by peptide mass fingerprinting or tandem mass spectrometry. Hence, this latter approach (also called "bottom-up" proteomics) uses identification at the peptide level to infer the existence of proteins.

Whole protein mass analysis is primarily conducted using either time-of-flight (TOF) MS, or Fourier transform ion cyclotron resonance (FT-ICR). These two types of instrument are preferable here because of their wide mass range, and in the case of FT-ICR, its high mass accuracy. Mass analysis of proteolytic peptides is a much more popular method of protein characterisation, as cheaper instrument designs can be used for characterisation. Additionally, sample preparation is easier once whole proteins have been digested into smaller peptide fragments. The most widely used instrument for peptide mass analysis are the MALDI time-of-flight instruments as they permit the acquisition of peptide mass fingerprints (PMFs) at high pace (1 PMF can be analysed in approx. 10 sec). Multiple stage quadrupole-time-of-flight and the quadrupole ion trap also find use in this application.

Protein and Peptide Fractionation Coupled with Mass Spectrometry

Proteins of interest to biological researchers are usually part of a very complex mixture of other proteins and molecules that co-exist in the biological medium. This presents two significant problems. First, the two ionization techniques used for large molecules only work well when the mixture contains roughly equal amounts of constituents, while in biological samples, different proteins tend to be present in widely differing amounts. If such a mixture is ionized using electrospray or MALDI, the more abundant species have a tendency to "drown" or suppress signals from less abundant ones. The second problem is that the mass spectrum from a complex mixture is very difficult to interpret because of the overwhelming number of mixture components. This is exacerbated by the fact that enzymatic digestion of a protein gives rise to a large number of peptide products.

To contend with this problem, two methods are widely used to fractionate proteins, or their peptide products from an enzymatic digestion. The first method fractionates whole proteins and is called two-dimensional gel electrophoresis. The second method, high performance liquid chromatography is used to fractionate peptides after enzymatic digestion. In some situations, it may be necessary to combine both of these techniques.

Gel spots identified on a 2D Gel are usually attributable to one protein. If the identity of the protein is desired, usually the method of in-gel digestion is applied, where the protein spot of interest is excised, and digested proteolytically. The peptide masses resulting from the digestion can be determined by mass spectrometry using peptide mass fingerprinting. If this information does not allow unequivocal identification of the protein, its peptides can be subject to tandem mass spectrometry for de novo sequencing.

Characterisation of protein mixtures using HPLC/MS is also called *shotgun proteomics* and *mudpit*. A peptide mixture that results from digestion of a protein mixture is fractionated by one or two steps of liquid chromatography. The eluent from the chromatography stage can be either directly introduced to the mass spectrometre through electrospray ionization, or laid down on a series of small spots for later mass analysis using MALDI.

Protein Identification

There are two main ways MS is used to identify proteins. Peptide mass fingerprinting (mentioned in the previous section) uses the masses of proteolytic peptides as input to a search of a database of predicted masses that would arise from digestion of a list of known proteins. If a protein sequence in the reference list gives rise to a significant number of predicted masses that match the experimental values, there is some evidence that this protein was present in the original sample.

Tandem MS is becoming a more popular experimental method for identifying proteins. Collision-induced dissociation is used in mainstream applications to generate a set of fragments from a specific peptide ion. The fragmentation process primarily gives rise to cleavage products that break along peptide bonds. Because of this simplicity in fragmentation, it is possible to use the observed fragment masses to match with a database of predicted masses for one of many given peptide sequences. Tandem MS of whole protein ions has been investigated recently using electron capture dissociation and has demonstrated extensive sequence information in principle but is not in common practice. This is sometimes referred to as the "top-down" approach in that it involves starting with the whole mass and then pulling it apart rather than starting with pieces (proteolytic fragments) and piecing the protein back together using de novo repeat detection (bottom-up).

De Novo *(Peptide) Sequencing*

De novo (peptide) sequencing for mass spectrometry is typically performed without prior knowledge of the amino acid sequence. It is the process of assigning amino acids from peptide fragment masses of a protein. De novo sequencing has proven successful for confirming and expanding upon results from database searches.

As de novo sequencing is based on mass and some amino acids have identical masses (e.g. leucine and isoleucine), accurate manual sequencing can be difficult. Therefore it may be necessary to utilise a sequence homology search application to work in tandem between a database search and de novo sequencing to address this inherent limitation.

Database searching has the advantage of quickly identifying sequences, provided they have already been documented in a database. Other inherent limitations of database searching include:

- Sequence modifications/mutations: some database searches do not adequately account for alterations to the 'documented' sequence, thus can miss valuable information.
- The unknown: if a sequence is not documented, it will not be found
- False positives
- Incomplete and corrupted data: a common, unnoticed problem

Annotation peptide spectral library can also be used as a reference for protein/peptide identification. It offers the unique strength of reduced search space and increased specificity. The limitations include the following:

- Spectra not included in the library will not be identified.
- Spectra collected from different type of mass spectrometres can have quite distinct features.
- Reference spectra in the library may contain noise peak, which may lead to false positive identifications.

A number of different algorithmic approaches have been described to identify peptides and proteins from tandem mass spectrometry (MS/MS), peptide de novo sequencing and sequence tag-based searching.

Protein Quantitation

Several recent methods allow for the quantitation of proteins by mass spectrometry (quantitative proteomics). Typically, stable (e.g. non-radioactive) heavier isotopes of carbon (^{13}C) or nitrogen (^{15}N) are incorporated into one sample while the other one is labelled with corresponding light isotopes (e.g. ^{12}C and ^{14}N). The two samples are mixed before the analysis. Peptides derived from the different samples can be distinguished due to their mass difference. The ratio of their peak intensities corresponds to the relative abundance ratio of the peptides (and proteins). The most popular methods for isotope labelling are SILAC (stable isotope labelling by amino acids in cell culture), trypsin-catalysed ^{18}O labelling, ICAT (isotope coded affinity tagging), iTRAQ (isobaric tags for relative and absolute quantitation).

"Semi-quantitative" mass spectrometry can be performed without labelling of samples. Typically, this is done with MALDI analysis (in linear mode). The peak intensity, or the peak area, from individual molecules (typically proteins) is here correlated to the amount of protein in the sample. However, the individual signal depends on the primary structure of the protein, on the complexity of the sample, and on the settings of the instrument. Other types of "label-free" quantitative mass spectrometry, uses the spectral counts (or peptide counts) of digested proteins as a means for determining relative protein amounts.

Protein Structure

Characteristics indicative of the 3-dimensional structure of proteins can be probed with mass spectrometry in various ways. By using chemical crosslinking to couple parts of the protein that are close in space, but far apart in sequence, information about the overall structure can be inferred. By following the exchange of amide protons with deuterium from the solvent, it is possible to probe the solvent accessibility of various parts of the protein. Another interesting avenue in protein structural studies is laser-induced covalent labelling. In this technique, solvent-exposed sites of the protein are modified by hydroxyl radicals. Its combination with rapid mixing has been used in protein folding studies.

Practical Applications

Biomarkers

The FDA defines a biomarker as, "A characteristic that is objectively measured and evaluated as an indicator of normal biologic processes, pathogenic processes, or pharmacologic responses to a therapeutic intervention". Mass spectrometry enables large-scale discovery of candidates for biomarkers.

Proteogenomics

In what is now commonly referred to as proteogenomics, proteomic technologies such as mass spectrometry are used for improving gene and protein annotations. Parallel analysis of the genome and the proteome facilitates discovery of post-translational modifications and proteolytic events, especially when comparing multiple species.

Glycan Analysis

Mass spectrometry (MS), with its low sample requirement and high sensitivity, has been predominantly used in glycobiology for characterisation and elucidation of glycan structures. Mass spectrometry provides a complementary method to HPLC for the analysis of glycans. Intact glycans may be detected directly as singly charged ions by matrix-assisted laser desorption/ionization mass spectrometry (MALDI-MS) or, following permethylation or peracetylation, by fast atom bombardment mass spectrometry (FAB-MS). Electrospray ionization mass spectrometry (ESI-MS) also gives good signals for the smaller glycans. Various free and commercial software are now available which interpret MS data and aid in Glycan structure characterisation.

Space Exploration

As a standard method for analysis, mass spectrometres have reached other planets and moons. Two were taken to Mars by the

Viking program. In early 2005 the Cassini–Huygens mission delivered a specialised GC-MS instrument aboard the Huygens probe through the atmosphere of Titan, the largest moon of the planet Saturn. This instrument analysed atmospheric samples along its descent trajectory and was able to vapourise and analyse samples of Titan's frozen, hydrocarbon covered surface once the probe had landed. These measurements compare the abundance of isotope(s) of each particle comparatively to earth's natural abundance. Also onboard the Cassini–Huygens spacecraft is an ion and neutral mass spectrometre which has been taking measurements of Titan's atmospheric composition as well as the composition of Enceladus' plumes. A Thermal and Evolved Gas Analyser mass spectrometre was carried by the Mars Phoenix Lander launched in 2007.

Mass spectrometres are also widely used in space missions to measure the composition of plasmas. For example, the Cassini spacecraft carries the Cassini Plasma Spectrometre (CAPS), which measures the mass of ions in Saturn's magnetosphere.

Respired Gas Monitor

Mass spectrometres were used in hospitals for respiratory gas analysis beginning around 1975 through the end of the century. Some are probably still in use but none are currently being manufactured. Found mostly in the operating room, they were a part of a complex system, in which respired gas samples from patients undergoing anesthesia were drawn into the instrument through a valve mechanism designed to sequentially connect up to 32 rooms to the mass spectrometre. A computer directed all operations of the system. The data collected from the mass spectrometre was delivered to the individual rooms for the anesthesiologist to use. The uniqueness of this magnetic sector mass spectrometre may have been the fact that a plane of detectors, each purposely positioned to collect all of the ion species expected to be in the samples, allowed the instrument to simultaneously report all of the gases respired by the patient. Although the mass range was limited to slightly over 120 u, fragmentation of some of the heavier molecules negated the need for a higher detection limit.

Nuclear Magnetic Resonance Spectroscopy

Nuclear magnetic resonance spectroscopy, most commonly known as NMR spectroscopy, is a research technique that exploits the magnetic properties of certain atomic nuclei to determine physical and chemical properties of atoms or the molecules in which they are contained. It relies on the phenomenon of nuclear magnetic resonance and can provide detailed information about the structure, dynamics, reaction state, and chemical environment of molecules.

Most frequently, NMR spectroscopy is used by chemists and biochemists to investigate the properties of organic molecules, though it is applicable to any kind of sample that contains nuclei possessing spin. Suitable samples range from small compounds analysed with 1-dimensional proton or carbon-13 NMR spectroscopy to large proteins or nucleic acids using 3 or 4-dimensional techniques. The impact of NMR spectroscopy on the sciences has been substantial because of the range of information and the diversity of samples, including solutions and solids.

Basic NMR Techniques

When placed in a magnetic field, NMR active nuclei (such as ^{1}H or ^{13}C) absorb electromagnetic radiation at a frequency characteristic of the isotope. The resonant frequency, energy of the absorption, and the intensity of the signal are proportional to the strength of the magnetic field. For example, in a 21 Tesla magnetic field, protons resonate at 900 MHz. It is common to refer to a 21 T magnet as a 900 MHz magnet, although different nuclei resonate at a different frequency at this field strength in proportion to their nuclear magnetic moments.

Chemical Shift

A spinning charge generates a magnetic field that results in a magnetic moment proportional to the spin. In the presence of an external magnetic field, two spin states exist (for a spin 1/2 nucleus): one spin up and one spin down, where one aligns with the magnetic field and the other opposes it. The difference in energy (ΔE) between the two spin states increases as the strength of the field increases, but this difference is usually very small, leading to the requirement for strong NMR magnets (1-20 T for modern NMR instruments). Irradiation of the sample with energy corresponding to the exact spin state separation of a specific set of nuclei will cause excitation of those set of nuclei in the lower energy state to the higher energy state.

For spin 1/2 nuclei, the energy difference between the two spin states at a given magnetic field strength are proportional to their magnetic moments. However, even if all protons have the same magnetic moments, they do not give resonant signals at the same field/frequency values. This is because this is dependent on the electrons surrounding the proton in covalent compounds. Upon application of an external magnetic field, these electrons move in response to the field and generate local magnetic fields that oppose the much stronger applied field. This local field thus "shields" the proton from the applied magnetic field, which must therefore be increased in order to achieve resonance (absorption of rf energy). Such increments are very small, usually in parts per million (ppm). The difference between 2.3487T and 2.3488T is therefore about 42ppm. However a frequency scale is commonly used to designate the NMR signals, even though the spectrometre may operate by sweeping the magnetic field, and thus the 42 ppm is 4200 Hz for a 100 MHz reference frequency (rf).

However given that the location of different NMR signals is dependent on the external magnetic field strength and the rf frequency, the signals are usually reported relative to a reference signal, usually that of TMS (tetramethylsilane). Additionally, since the distribution of NMR signals is field dependent, these frequencies are divided by the spectrometre frequency.

However since we are dividing Hz by MHz, the resulting number would be too small, and thus it is multiplied by a million. This operation therefore gives a locator number called the "chemical shift" with units of parts per million. To detect such small frequency differences the applied magnetic field must be constant throughout the sample volume. High resolution NMR spectrometres use shims to adjust the homogeneity of the magnetic field to parts per billion (ppb) in a volume of a few cubic centimetres. In general, chemical shifts for protons are highly predictable since the shifts are primarily determined by simpler shielding effects (electron density), but the chemical shifts for many heavier nuclei are more strongly influenced by other factors including excited states ("paramagnetic" contribution to shielding tensor).

The chemical shift provides information about the structure of the molecule. The conversion of the raw data to this information is called *assigning* the spectrum. For example, for the ^{1}H-NMR spectrum for ethanol (CH_3CH_2OH), one would expect signals at each of three specific chemical shifts: one for the CH_3 group, one for the CH_2 group and one for the OH group. A typical CH_3 group has a shift around

1 ppm, a CH_2 attached to an OH has a shift of around 4 ppm and an OH has a shift around 2–3 ppm depending on the solvent used.

Because of molecular motion at room temperature, the three methyl protons *average out* during the course of the NMR experiment (which typically requires a few ms). These protons become degenerate and form a peak at the same chemical shift.

The shape and area of peaks are indicators of chemical structure too. In the example above—the proton spectrum of ethanol—the CH_3 peak has three times the area as the OH peak. Similarly the CH_2 peak would be twice the area of the OH peak but only 2/3 the area of the CH_3 peak.

Software allows analysis of the size of peaks to understand how many protons give rise to the peak. This is known as integration—a mathematical process which calculates the area under a curve. The analyst must integrate the peak and not measure its height because the peaks also have *width*—and thus its size is dependent on its area not its height. However, it should be mentioned that the number of protons, or any other observed nucleus, is only proportional to the intensity, or the integral, of the NMR signal in the very simplest one-dimensional NMR experiments. In more elaborate experiments, for instance, experiments typically used to obtain carbon-13 NMR spectra, the integral of the signals depends on the relaxation rate of the nucleus, and its scalar and dipolar coupling constants. Very often these factors are poorly known - therefore, the integral of the NMR signal is very difficult to interpret in more complicated NMR experiments.

In nuclear magnetic resonance (NMR) spectroscopy, the chemical shift is the resonant frequency of a nucleus relative to a standard. Often the position and number of chemical shifts are diagnostic of the structure of a molecule. Chemical shifts are also used to describe signals in other forms of spectroscopy such as photoemission spectroscopy.

Some atomic nuclei possess a magnetic moment (nuclear spin), which gives rise to different energy levels and resonance frequencies in a magnetic field. The total magnetic field experienced by a nucleus includes local magnetic fields induced by currents of electrons in the molecular orbitals (note that electrons have a magnetic moment themselves). The electron distribution of the same type of nucleus (e.g. ^{1}H, ^{13}C, ^{15}N) usually varies according to the local geometry (binding partners, bond lengths, angles between bonds, ...), and with it the local magnetic field at each nucleus.

This is reflected in the spin energy levels (and resonance frequencies). The variations of nuclear magnetic resonance frequencies of the same kind of nucleus, due to variations in the electron distribution, are called the chemical shift. The size of the chemical shift is given with respect to a reference frequency or reference sample, usually a molecule with a barely distorted electron distribution.

Operating Frequency

The operating (or Larmor) frequency ω_0 of a magnet is calculated from the Larmor equation

$$\omega_0 = \gamma * B_0$$

where B_0 is the actual strength of the magnet in units like teslas or gauss, and γ is the gyromagnetic ratio of the nucleus being tested which is in turn calculated from its magnetic moment μ and spin number I with the nuclear magneton μ_N and the Planck constant h:

$$\gamma = \frac{\mu \mu_N}{hI}$$

Thus, the proton operating frequency for a 1 T magnet is calculated as:

$$\omega_0 = \gamma B_0 = \frac{2.79 \times 5.05 \times 10^{-27}\ \mathrm{J/T}}{6.62 \times 10^{-34}\ \mathrm{Js} \times (1/2)} \times 1\mathrm{T} = 42.5\mathrm{MHz}$$

Chemical Shift Referencing

Chemical shift δ is usually expressed in parts per million (ppm) by frequency, because it is calculated from:

$$\delta = \frac{\text{difference between a resonance frequency and that of a reference substance}}{\text{operating frequency of the spectrometer}}$$

Since the numerator is usually in hertz, and the denominator in megahertz, delta is expressed in ppm.

The detected frequencies (in Hz) for ^{1}H, ^{13}C, and ^{29}Si nuclei are usually referenced against TMS (tetramethylsilane) or DSS, which is assigned the chemical shift of zero. Other standard materials are used for setting the chemical shift for other nuclei. Thus, an NMR signal that absorbs at 300 Hz higher than does TMS at an applied frequency of 300 MHz has a chemical shift of:

$$\frac{300\text{Hz}}{300\times10^{6}\text{Hz}}=1\times10^{-6}=1\text{ppm}$$

Although the frequency depends on the applied field the chemical shift is independent of it. On the other hand the resolution of NMR will increase with applied magnetic field resulting in ever increasing chemical shift changes.

The Induced Magnetic Field

The electrons around a nucleus will circulate in a magnetic field and create a secondary induced magnetic field. This field opposes the applied field as stipulated by Lenz's law and atoms with higher induced fields (i.e., higher electron density) are therefore called *shielded*, relative to those with lower electron density. The chemical milieu of an atom can influence its electron density through the polar effect. Electron-donating alkyl groups, for example, lead to increased shielding while electron-withdrawing substituents such as nitro groups lead to *deshielding* of the nucleus. Not only substituents cause local induced fields. Bonding electrons can also lead to shielding and deshielding effects. A striking example of this are the pi bonds in benzene. Circular current through the hyperconjugated system causes a shielding effect at the molecule's centre and a deshielding effect at its edges. Trends in chemical shift are explained based on the degree of shielding or deshielding.

Nuclei are found to resonate in a wide range to the left (or more rare to the right) of the internal standard. When a signal is found with a higher chemical shift:

- the applied effective magnetic field is lower, if the resonance frequency is fixed, (as in old traditional CW spectrometres)
- the frequency is higher, when the applied magnetic field is static, (normal case in FT spectrometres)
- the nucleus is more deshielded
- the signal or shift is downfield or at low field or paramagnetic

Conversely a lower chemical shift is called a diamagnetic shift, and is upfield and more shielded.

Diamagnetic Shielding

In real molecules protons are surrounded by a cloud of charge due to adjacent bonds and atoms. In an applied magnetic field (B_0) electrons circulate and produce an induced field (B_i) which opposes the applied field. The effective field at the nucleus will be $B = B_0 - B_i$. The nucleus is said to be experiencing a diamagnetic shielding.

Factors Causing Chemical Shifts

Important factors influencing chemical shift are electron density, electronegativity of neighbouring groups and anisotropic induced magnetic field effects.

Electron density shields a nucleus from the external field. For example in proton NMR the electron-poor tropylium ion has its protons downfield at 9.17 ppm, those of the electron-rich cyclooctatetraenyl anion move upfield to 6.75 ppm and its dianion even more upfield to 5.56 ppm.

A nucleus in the vicinity of an electronegative atom experiences reduced electron density and the nucleus is therefore deshielded. In proton NMR of methyl halides (CH_3X) the chemical shift of the methyl protons increase in the order I < Br < Cl < F from 2.16 ppm to 4.26 ppm reflecting this trend. In carbon NMR the chemical shift of the carbon nuclei increase in the same order from around –10 ppm to 70 ppm. Also when the electronegative atom is removed further away the effect diminishes until it can be observed no longer.

Anisotropic induced magnetic field effects are the result of a local induced magnetic field experienced by a nucleus resulting from circulating electrons that can either be paramagnetic when it is parallel to the applied field or diamagnetic when it is opposed to it. It is observed in alkenes where the double bond is oriented perpendicular to the external field with pi electrons likewise circulating at right angles. The induced magnetic field lines are parallel to the external field at the location of the alkene protons which therefore shift downfield to a 4.5 ppm to 7.5 ppm range. The three-dimensional space where a nucleus experiences diamagnetic shift is called the shielding zone with a cone-like shape aligned with the external field.

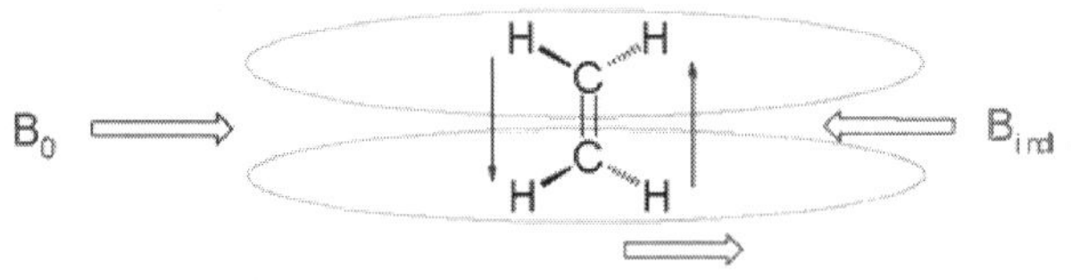

Figure: The protons in aromatic compounds are shifted downfield even further with a signal for benzene at 7.73 ppm as a consequence of a diamagnetic ring current.

Alkyne protons by contrast resonate at high field in a 2–3 ppm range. For alkynes the most effective orientation is the external field in parallel with electrons circulation around the triple bond. In this way the acetylenic protons are located in the cone-shaped shielding zone hence the upfield shift.

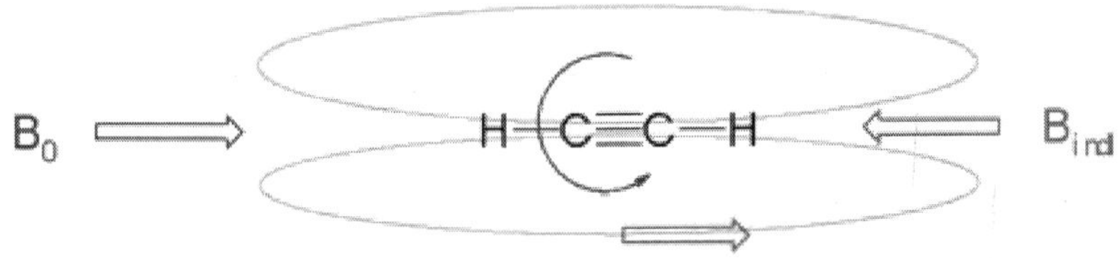

Magnetic Properties of Most Common Nuclei

^{1}H and ^{13}C aren't the only nuclei susceptible to NMR experiments. A number of different nuclei can also be detected, although the use of such techniques is generally rare due to small relative sensitivities in NMR experiments (compared to ^{1}H) of the nuclei in question, the other factor for rare use being their slender representation in nature/ organic compounds.

Isotope	*Occurrence in nature (%)*	*spin number l*	*Magnetic moment μ*	*Electric quadrupole moment ($e\times10^{?24}$ cm^2)*	*Operating frequency at 7 T (MHz)*	*Relative sensitivity*
^{1}H	99.984	1/2	2.79628		300.13	1
^{2}H	0.016	1	0.85739	$2.8 \times 10^{?3}$	46.07	0.0964
^{10}B	18.8	3	1.8005	$7.4 \times 10^{?2}$	32.25	0.0199
^{11}B	81.2	3/2	2.6880	$2.6 \times 10^{?2}$	96.29	0.165
^{12}C	98.9	0				
^{13}C	1.1	1/2	0.70220		75.47	0.0159
^{14}N	99.64	1	0.40358	$7.1 \times 10^{?2}$	21.68	0.00101
^{15}N	0.37	1/2	?0.28304		30.41	0.00104
^{16}O	99.76	0				
^{17}O	0.0317	5/2	?1.8930	$?4.0 \times 10^{?3}$	40.69	0.0291
^{19}F	100	1/2	2.6273		282.40	0.834
^{28}Si	92.28	0				
^{29}Si	4.70	1/2	?0.55548		59.63	0.0785
^{31}P	100	1/2	1.1205		121.49	0.0664
^{35}Cl	75.4	3/2	0.92091	$?7.9 \times 10^{?2}$	29.41	0.0047
^{37}Cl	24.6	3/2	0.68330	$?6.2 \times 10^{?2}$	24.48	0.0027

Magnetic properties of common nuclei

^{1}H, ^{13}C, ^{15}N, ^{19}F and ^{31}P are the five nuclei that have the greatest importance in NMR experiments:

- ^{1}H because of high sensitivity and vast occurrence in organic compounds
- ^{13}C because of being the key component of all organic compounds despite occurring at a low abundance (1.1%) compared to the major isotope of carbon ^{12}C, which has a spin of 0 and therefore is NMR inactive.

- ^{15}N because of being a key component of important biomolecules such as proteins and DNA
- ^{19}F because of high relative sensitivity

•^{31}P because of frequent occurrence in organic compounds and moderate relative sensitivity

Other Chemical Shifts

The related Knight shift (first reported in 1949) is observed with pure metals. The NMR chemical shift in its present day meaning first appeared in journals in 1950. Chemical shifts with a different meaning appear in X-ray photoelectron spectroscopy as the shift in atomic core-level energy due to a specific chemical environment. The term is also used in Mössbauer spectroscopy, where similarly to NMR it refers to a shift in peak position due to the local chemical bonding environment. As is the case for NMR the chemical shift reflects the electron density at the atomic nucleus.

J-coupling

Multiplicity	*Intensity Ratio*
Singlet (s)	1
Doublet (d)	1:1
Triplet (t)	1:2:1
Quartet (q)	1:3:3:1
Quintet	1:4:6:4:1
Sextet	1:5:10:10:5:1
Septet	1:6:15:20:15:6:1

Some of the most useful information for structure determination in a one-dimensional NMR spectrum comes from J-coupling or scalar coupling (a special case of spin-spin coupling) between NMR active nuclei. This coupling arises from the interaction of different spin states through the chemical bonds of a molecule and results in the splitting of NMR signals. These splitting patterns can be complex or simple and, likewise, can be straightforwardly interpretable or deceptive. This coupling provides detailed insight into the connectivity of atoms in a molecule.

Coupling to n equivalent (spin ½) nuclei splits the signal into a n+1 multiplet with intensity ratios following Pascal's triangle as described on the right. Coupling to additional spins will lead to further splittings of each component of the multiplet e.g. coupling to two different spin ½ nuclei with significantly different coupling constants will lead to a *doublet of doublets* (abbreviation: dd). Note that coupling between nuclei that are chemically equivalent (that is, have the same chemical shift) has no effect of the NMR spectra and couplings between nuclei that are distant (usually more than 3 bonds apart for protons in flexible molecules) are usually too small to cause observable splittings. *Long-range* couplings over more than three bonds can often be observed in cyclic and aromatic compounds, leading to more complex splitting patterns.

For example, in the proton spectrum for ethanol described above, the CH_3 group is split into a *triplet* with an intensity ratio of 1:2:1 by the two neighbouring CH_2 protons. Similarly, the CH_2 is split into a *quartet* with an intensity ratio of 1:3:3:1 by the three neighbouring CH_3 protons. In principle, the two CH_2 protons would also be split again into a *doublet* to form a *doublet of quartets* by the hydroxyl proton, but intermolecular exchange of the acidic hydroxyl proton often results in a loss of coupling information.

Coupling to any spin ½ nuclei such as phosphorus-31 or fluorine-19 works in this fashion (although the magnitudes of the coupling constants may be very different). But the splitting patterns differ from those described above for nuclei with spin greater than ½ because the spin quantum number has more than two possible values. For instance, coupling to deuterium (a spin 1 nucleus) splits the signal into a *1:1:1 triplet* because the spin 1 has three spin states. Similarly, a spin 3/2 nucleus splits a signal into a *1:1:1:1 quartet* and so on.

Coupling combined with the chemical shift (and the integration for protons) tells us not only about the chemical environment of the nuclei, but also the number of *neighbouring* NMR active nuclei within the molecule. In more complex spectra with multiple peaks at similar chemical shifts or in spectra of nuclei other than hydrogen, coupling is often the only way to distinguish different nuclei.

Second-Order (or Strong) Coupling

The above description assumes that the coupling constant is small in comparison with the difference in NMR frequencies between the inequivalent spins. If the shift separation decreases (or the coupling strength increases), the multiplet intensity patterns are first distorted, and then become more complex and less easily analysed (especially if more than two spins are involved). Intensification of some peaks in a multiplet is achieved at the expense of the remainder, which sometimes almost disappear in the background noise, although the integrated area under the peaks remains constant. In most high-field NMR, however, the distortions are usually modest and the characteristic distortions (*roofing*) can in fact help to identify related peaks.

Second-order effects decrease as the frequency difference between multiplets increases, so that high-field (i.e. high-frequency) NMR spectra display less distortion than lower frequency spectra. Early spectra at 60 MHz were more prone to distortion than spectra from later machines typically operating at frequencies at 200 MHz or above.

Magnetic Inequivalence

More subtle effects can occur if chemically equivalent spins (i.e., nuclei related by symmetry and so having the same NMR frequency) have different coupling relationships to external spins. Spins that are chemically equivalent but are not indistinguishable (based on their coupling relationships) are termed magnetically inequivalent. For example, the 4 H sites of 1,2-dichlorobenzene divide into two chemically equivalent pairs by symmetry, but an individual member of one of the pairs has different couplings to the spins making up the other pair. Magnetic inequivalence can lead to highly complex spectra which can only be analysed by computational modelling. Such effects are more common in NMR spectra of aromatic and other non-flexible systems, while conformational averaging about C-C bonds in flexible molecules tends to equalize the couplings between protons on adjacent carbons, reducing problems with magnetic inequivalence.

Correlation Spectroscopy

Correlation spectroscopy is one of several types of two-dimensional nuclear magnetic resonance (NMR) spectroscopy or 2D-NMR. This type of NMR experiment is best known by its acronym, COSY. Other types of two-dimensional NMR include J-spectroscopy, exchange spectroscopy (EXSY), Nuclear Overhauser effect spectroscopy (NOESY), total correlation spectroscopy (TOCSY) and heteronuclear correlation experiments, such as HSQC, HMQC, and HMBC. Two-dimensional NMR spectra provide more information about a molecule than one-dimensional NMR spectra and are especially useful in determining the structure of a molecule, particularly for molecules that are too complicated to work with using one-dimensional NMR. The first two-dimensional experiment, COSY, was proposed by Jean Jeener, a professor at Université Libre de Bruxelles, in 1971. This experiment was later implemented by Walter P. Aue, Enrico Bartholdi and Richard R. Ernst, who published their work in 1976.

Solid-State Nuclear Magnetic Resonance

A variety of physical circumstances does not allow molecules to be studied in solution, and at the same time not by other spectroscopic techniques to an atomic level, either. In solid-phase media, such as crystals, microcrystalline powders, gels, anisotropic solutions, etc., it is in particular the dipolar coupling and chemical shift anisotropy that become dominant to the behaviour of the nuclear spin systems. In conventional solution-state NMR spectroscopy, these additional interactions would lead to a significant broadening of spectral lines. A variety of techniques allows to establish high-resolution conditions, that can, at least for ^{13}C spectra, be comparable to solution-state NMR spectra.

Two important concepts for high-resolution solid-state NMR spectroscopy are the limitation of possible molecular orientation by sample orientation, and the reduction of anisotropic nuclear magnetic interactions by sample spinning. Of the latter approach, fast spinning around the magic angle is a very prominent method, when the system comprises spin 1/2 nuclei. A number of intermediate techniques, with samples of partial alignment or reduced mobility, is currently being used in NMR spectroscopy.

Applications in which solid-state NMR effects occur are often related to structure investigations on membrane proteins, protein fibrils or all kinds of polymers, and chemical analysis in inorganic chemistry, but also include "exotic" applications like the plant leaves and fuel cells.

Solid-state NMR (SSNMR) spectroscopy is a kind of nuclear magnetic resonance (NMR) spectroscopy, characterised by the presence of anisotropic (directionally dependent) interactions.

Basic concepts A spin interacts with a magnetic or an electric field. Spatial proximity and/or a chemical bond between two atoms can give rise to interactions between nuclei. In general, these interactions are orientation dependent. In media with no or little mobility (e.g. crystals, powders, large membrane vesicles, molecular aggregates), anisotropic interactions have a substantial influence on the behaviour of a system of nuclear spins. In contrast, in a classical liquid-state NMR experiment, Brownian motion leads to an averaging of anisotropic interactions. In such cases, these interactions can be neglected on the time-scale of the NMR experiment.

Examples of anisotropic nuclear interactions Two directionally dependent interactions commonly found in solid-state NMR are the *chemical shift anisotropy* (CSA) and the internuclear *dipolar coupling*. Many more such interactions exist, such as the anisotropic J-coupling in NMR, or in related fields, such as the *g*-tensor in electron spin resonance. In mathematical terms, all these interactions can be described using the same formalism.

Experimental background Anisotropic interactions modify the nuclear spin energy levels (and hence the resonance frequency) of all sites in a molecule, and often contribute to a line-broadening effect in NMR spectra. However, there is a range of situations when their presence can either not be avoided, or is even particularly desired, as they encode structural parameters, such as orientation information, on the molecule of interest.

High-resolution conditions in solids (in a wider sense) can be established using magic angle spinning (MAS), macroscopic sample orientation, combinations of both of these techniques, enhancement of mobility by highly viscous sample conditions, and a variety of radio frequency (RF) irradiation patterns. While the latter allows decoupling of interactions in spin space, the others facilitate averaging of interactions in real space. In addition, line-broadening effects from microscopic inhomogeneities can be reduced by appropriate methods of sample preparation.

Under decoupling conditions, isotropic interactions can report on the local structure, e.g. by the isotropic chemical shift. In addition, decoupled interactions can be selectively re-introduced ("recoupling"), and used, for example, for controlled de-phasing or transfer of polarization to derive a number of structural parameters.

Solid-state NMR line widths The residual line width (full width at half max) of ^{13}C nuclei under MAS conditions at 5–15 kHz spinning rate is typically in the order of 0.5–2 ppm, and may be comparable to solution-state NMR conditions. Even at MAS rates of 20 kHz and above, however, non linear groups (not a straight line) of the same nuclei linked via the homonuclear dipolar interactions can only be suppressed partially, leading to line widths of 0.5 ppm and above, which is considerably more than in optimal solution state NMR conditions. Other interactions such as the quadrupolar interaction can lead to line widths of thousands of ppm due to the strength of the interaction. The first-order quadrupolar broadening is largely suppressed by sufficiently fast MAS, but the second-order quadrupolar broadening has a different angular dependence and cannot be removed by spinning at one angle alone. Ways to achieve isotropic lineshapes for quadrupolar nuclei include spinning at two angles simultaneously (DOR), sequentially (DAS), or through refocusing the second-order quadrupolar interaction with a two-dimensional experiment such as MQMAS or STMAS.

Anisotropic interactions in solution-state NMR From the perspective of solution-state NMR, it can be desirable to reduce motional averaging of dipolar interactions by alignment media. The order of magnitude of these residual dipolar couplings (RDCs) are typically of only a few rad/Hz, but do not destroy high-resolution conditions, and provide a pool of information, in particular on the orientation of molecular domains with respect to each other.

Dipolar truncation The dipolar coupling between two nuclei is inversely proportional to the cube of their distance. This has the effect that the polarization transfer mediated by the dipolar interaction is cut off in the presence of a third nucleus (all of the same kind, e.g. ^{13}C) close to one of these nuclei. This effect is commonly referred to as dipolar truncation. It has been one of the major obstacles in efficient extraction of internuclear distances, which are crucial in the structural analysis of biomolecular structure. By means of labelling schemes or pulse sequences, however, it has become possible to circumvent this problem in a number of ways.

Nuclear Spin Interactions in the Solid Phase

Chemical Shielding

The chemical shielding is a local property of each nucleus, and depends on the external magnetic field.

Specifically, the external magnetic field induces currents of the electrons in molecular orbitals. These induced currents create local magnetic fields that often vary across the entire molecular framework such that nuclei in distinct molecular environments usually experience unique local fields from this effect.

Under sufficiently fast magic angle spinning, or in solution-state NMR, the directionally dependent character of the chemical shielding is removed, leaving the isotropic chemical shift.

J-Coupling

The J-coupling or indirect nuclear spin-spin coupling (sometimes also called "scalar" coupling despite the fact that J is a tensor quantity) describes the interaction of nuclear spins through chemical bonds.

Dipolar Coupling

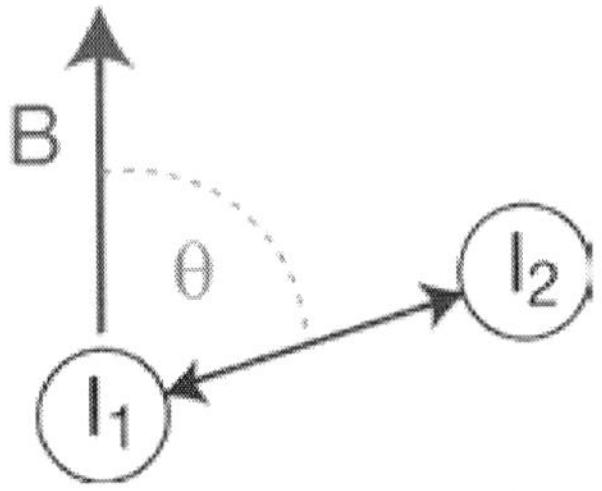

Figure: Dipolar coupling vectors

Nuclear spins exhibit a dipole moment, which interacts with the dipole moment of other nuclei (dipolar coupling). The magnitude of the interaction is dependent on the spin species, the internuclear distance, and the orientation of the vector connecting the two nuclear spins with respect to the external magnetic field *B*. The maximum dipolar coupling is given by the dipolar coupling constant *d*,

$$d = \frac{\hbar \mu_0}{4\pi} \frac{\gamma_1 \gamma_2}{r^3}$$

where r is the distance between the nuclei, and γ_1 and γ_2 are the gyromagnetic ratios of the nuclei. In a strong magnetic field, the dipolar coupling depends on the orientation of the internuclear vector with the external magnetic field by

$$D \propto 3\cos^2\theta - 1 \cdot$$

Consequently, two nuclei with a dipolar coupling vector at an angle of θ_m=54.7° to a strong external magnetic field, which is the angle where D becomes zero, have zero dipolar coupling. θ_m is called the magic angle. One technique for removing dipolar couplings, at least to some extent, is magic angle spinning.

Quadrupolar Interaction

Nuclei with a spin greater than one-half have a non spherical charge distribution. This is known as a quadrupolar nucleus. A non spherical charge distribution can interact with an electric field gradient caused by some form of non-symmetry (e.g. in a trigonal bonding atom there are electrons around it in a plane, but not above or below it) to produce a change in the energy level in addition to the Zeeman effect. The quadrupolar interaction is the largest interaction in NMR apart from the Zeeman interaction and they can even become comparable in size. Due to the interaction being so large it can not be treated to just the first order, like most of the other interactions. This means you have a first and second order interaction, which can be treated separately. The first order interaction has an angular dependency with respect to the magnetic field of $(3\cos^2\theta - 1)$ (the P2 Legendre polynomial), this means that if you spin the sample at $\theta = \arctan\sqrt{2}$ (~54.74°) you can average out the first order interaction over one rotor period (all other interactions apart from Zeeman, Chemical shift, paramagnetic and J coupling also have this angular dependency). However, the second order interaction depends on the P4 Legendre polynomial, which has zero points at 30.6° and 70.1°. These can be taken advantage of by either using DOR (DOuble angle Rotation) where you spin at two angles at the same time, or DAS (Double Angle Spinning) where you switch quickly between the two angles. Specialised hardware (probe) has been developed for such experiments. A revolutionary advance is Lucio Frydman's multiple quantum magic angle spinning (MQMAS) NMR in 1995 and it has become a routine method for obtaining high resolution solid-state NMR spectra of quadrupolar nuclei. A similar method to MQMAS is satellite transition magic angle spinning (STMAS) NMR proposed by Zhehong Gan in 2000.

Modern Solid-State NMR Spectroscopy

Basic Example

A fundamental RF pulse sequence and building-block in most solid-state NMR experiments starts with cross-polarization (CP) [Waugh *et al.*]. It can be used to enhance the signal of nuclei with a low gyromagnetic ratio (e.g. ^{13}C, ^{15}N) by magnetization transfer from nuclei with a high gyromagnetic ratio (e.g. ^{1}H), or as spectral editing method (e.g. directed $^{15}N \rightarrow ^{13}C$ CP in protein spectroscopy). To establish magnetization transfer, the RF pulses applied on the two frequency channels must fulfill the Hartmann–Hahn condition [Hartmann, 1962]. Under MAS, this condition defines a relationship between the voltage through the RF coil and the rate of sample rotation. Experimental optimization of such conditions is one of the routine tasks in performing a (solid-state) NMR experiment.

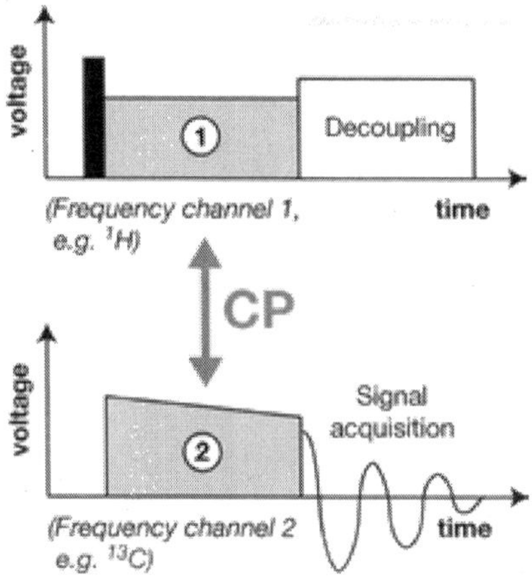

Figure: CP pulse sequence

CP-MAS is a basic building block of most pulse sequences in solid-state NMR spectroscopy. Given its importance, a pulse sequence employing direct excitation of ^1H spin polarization, followed by CP transfer to and signal detection of ^{13}C, ^{15}N) or similar nuclei, is itself often referred to as *CP experiment*, or, in conjunction with MAS, as *CP-MAS* [Schaefer and Stejskal, 1976]. It is the typical starting point of an investigation using solid-state NMR spectroscopy.

Decoupling

Nuclear spin interactions must be removed (decoupled) to increase the resolution of NMR spectra and isolate spin systems. A technique that can substantially reduce or remove the chemical shift anisotropy, the dipolar coupling is *sample rotation* (most commonly magic angle spinning, but also off-magic angle spinning).

Homonuclear RF decoupling decouples spin interactions of nuclei that are the same as those being detected. *Heteronuclear RF decoupling* decouples spin interactions of other nuclei.

Recoupling

Although the broadened lines are often not desired, dipolar couplings between atoms in the crystal lattice can also provide very useful information. Dipolar coupling are distance dependent, and so they may be used to calculate interatomic distances in isotopically labelled molecules.

Because most dipolar interactions are removed by sample spinning, recoupling experiments are needed to re-introduce desired dipolar couplings so they can be measured.

An example of a recoupling experiment is the Rotational Echo DOuble Resonance (REDOR) experiment which also can be the basis of an NMR crystallographic study of e.g. an amorphous solid.

Membrane proteins and amyloid fibrils, the latter related to Alzheimer's disease and Parkinson's disease, are two examples of application where solid-state NMR spectroscopy complements solution-state NMR spectroscopy and beam diffraction methods (e.g. X-ray crystallography, electron microscopy).

Solid-state NMR spectroscopy serves as an analysis tool in organic and inorganic chemistry. SSNMR is also a valuable tool to study local dynamics, kinetics, and thermodynamics of a variety of systems.

Biomolecular NMR Spectroscopy

Proteins

Much of the innovation within NMR spectroscopy has been within the field of protein NMR spectroscopy, an important technique in structural biology. A common goal of these investigations is to obtain high resolution 3-dimensional structures of the protein, similar to what can be achieved by X-ray crystallography. In contrast to X-ray crystallography, NMR spectroscopy is usually limited to proteins smaller than 35 kDa, although larger structures have been solved. NMR spectroscopy is often the only way to obtain high resolution information on partially or wholly intrinsically unstructured proteins. It is now a common tool for the determination of Conformation Activity Relationships where the structure before and after interaction with, for example, a drug candidate is compared to its known biochemical activity. Proteins are orders of magnitude larger than the small organic molecules discussed earlier in this chapter, but the basic NMR techniques and some of the NMR theory also applies. Because of the much higher number of atoms present in a protein molecule in comparison with a small organic compound, the basic 1D spectra become crowded with overlapping signals to an extent where direct spectra analysis becomes untenable. Therefore, multidimensional (2, 3 or 4D) experiments have been devised to deal with this problem. To facilitate these experiments, it is desirable to isotopically label the protein with ^{13}C and ^{15}N because the predominant naturally occurring isotope ^{12}C is not NMR-active, whereas the nuclear quadrupole moment of the predominant naturally occurring ^{14}N isotope prevents high resolution information to be obtained from this nitrogen isotope. The most important method used for structure determination of proteins utilises NOE experiments to measure distances between pairs of atoms within the molecule. Subsequently, the obtained distances are used to generate a 3D structure of the molecule by solving a distance geometry problem.

Nuclear magnetic resonance spectroscopy of proteins (usually abbreviated protein NMR) is a field of structural biology in which NMR spectroscopy is used to obtain information about the structure and dynamics of proteins. The field was pioneered by Richard R. Ernst and Kurt Wüthrich, among others. Protein NMR techniques are continually being used and improved in both academia and the biotech industry. Structure determination by NMR spectroscopy usually consists of several following phases, each using a separate set of highly specialised techniques. The sample is prepared, resonances are assigned, restraints are generated and a structure is calculated and validated.

Sample Preparation

Protein nuclear magnetic resonance is performed on aqueous samples of highly purified protein. Usually the sample consist of between 300 and 600 microlitres with a protein concentration in the range 0.1 – 3 millimolar. The source of the protein can be either natural or produced in an expression system using recombinant DNA techniques through genetic engineering. Recombinantly expressed proteins are usually easier to produce in sufficient quantity, and makes isotopic labelling possible. The purified protein is usually dissolved in a buffer solution and adjusted to the desired solvent conditions. The NMR sample is prepared in a thin walled glass tube.

Data Collection

Protein NMR utilises multidimensional nuclear magnetic resonance experiments to obtain information about the protein. Ideally, each distinct nucleus in the molecule experiences a distinct chemical environment and thus has a distinct chemical shift by which it can be recognized. However, in large molecules such as proteins the number of resonances can typically be several thousand and a one-dimensional spectrum inevitably has incidental overlaps. Therefore multidimensional experiments are performed which correlate the frequencies of distinct nuclei. The additional dimensions decrease the chance of overlap and have a larger information content since they correlate signals from nuclei within a specific part of the molecule.

Magnetization is transferred into the sample using pulses of electromagnetic (radiofrequency) energy and between nuclei using delays; the process is described with so-called pulse sequences. Pulse sequences allow the experimenter to investigate and select specific types of connections between nuclei. The array of nuclear magnetic resonance experiments used on proteins fall in two main categories — one where magnetization is transferred through the chemical bonds, and one where the transfer is through space, irrespective of the bonding structure. The first category is used to assign the different chemical shifts to a specific nucleus, and the second is primarily used to generate the distance restraints used in the structure calculation, and in the assignment with unlabelled protein.

Depending on the concentration of the sample, on the magnetic field of the spectrometre, and on the type of experiment, a single multidimensional nuclear magnetic resonance experiment on a protein sample may take hours or even several days to obtain suitable signal-to-noise ratio through signal averaging, and to allow for sufficient evolution of magnetization transfer through the various dimensions of the experiment. Other things being equal, higher-dimensional experiments will take longer than lower-dimensional experiments.

Typically the first experiment to be measured with an isotope-labelled protein is a 2D heteronuclear single quantum correlation (HSQC) spectrum where "heteronuclear" refers to nuclei other than 1H. In theory the heteronuclear single quantum correlation has one peak for each H bound to a heteronucleus. Thus in the 15N-HSQC one signal is expected for each amino acid residue with the exception of proline which has no amide-hydrogen due to the cyclic nature of its backbone.

Tryptophan and certain other residues with N-containing sidechains also give rise to additional signals. The 15N-HSQC is often referred to as the fingerprint of a protein because each protein has a unique pattern of signal positions. Analysis of the 15N-HSQC allows researchers to evaluate whether the expected number of peaks is present and thus to identify possible problems due to multiple conformations or sample heterogeneity. The relatively quick heteronuclear single quantum correlation experiment helps determine the feasibility of doing subsequent longer, more expensive, and more elaborate experiments. It is not possible to assign peaks to specific atoms from the heteronuclear single quantum correlation alone.

Resonance Assignment

In order to analyse the nuclear magnetic resonance data, it is important to get a resonance assignment for the protein. That is to find out which chemical shift corresponds to which atom. This is typically achieved by sequential walking using information derived from several different types of NMR experiment. The exact procedure depends on whether the protein is isotopically labelled or not, since a lot of the assignment experiments depend on carbon-13 and nitrogen-15.

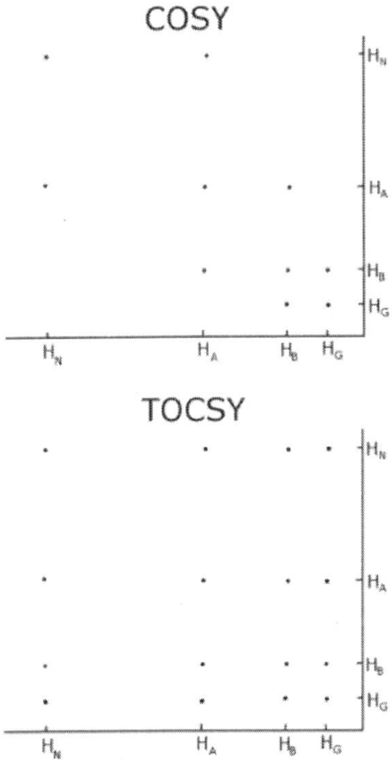

Figure: Comparison of a COSY and TOCSY 2D spectra for an amino acid like glutamate or methionine. The TOCSY shows off diagonal crosspeaks between all protons in the spectrum, but the COSY only has crosspeaks between neighbours.

Homonuclear Nuclear Magnetic Resonance

With unlabelled protein the usual procedure is to record a set of two dimensional homonuclear nuclear magnetic resonance experiments through correlation spectroscopy (COSY), of which several types include conventional correlation spectroscopy, *total correlation* spectroscopy (TOCSY) and nuclear Overhauser effect spectroscopy (NOESY). A two-dimensional nuclear magnetic resonance experiment produces a two-dimensional spectrum. The units of both axes are chemical shifts. The COSY and TOCSY transfer magnetization through the chemical bonds between adjacent protons.

The conventional correlation spectroscopy experiment is only able to transfer magnetization between protons on adjacent atoms, whereas in the total correlation spectroscopy experiment the protons are able to relay the magnetization, so it is transferred among all the protons that are connected by adjacent atoms. Thus in a conventional correlation spectroscopy, an alpha proton transfers magnetization to the beta protons, the beta protons transfers to the alpha and gamma protons, if any are present, then the gamma proton transfers to the beta and the delta protons, and the process continues.

In total correlation spectroscopy, the alpha and all the other protons are able to transfer magnetization to the beta, gamma, delta, epsilon if they are connected by a continuous chain of protons. The continuous chain of protons are the sidechain of the individual amino acids. Thus these two experiments are used to build so called spin systems, that is build a list of resonances of the chemical shift of the peptide proton, the alpha protons and all the protons from each residue's sidechain.

Which chemical shifts corresponds to which nuclei in the spin system is determined by the conventional correlation spectroscopy connectivities and the fact that different types of protons have characteristic chemical shifts. To connect the different spinsystems in a sequential order, the nuclear Overhauser effect spectroscopy experiment has to be used. Because this experiment transfers magnetization through space, it will show crosspeaks for all protons that are close in space regardless of whether they are in the same spin system or not. The neighbouring residues are inherently close in space, so the assignments can be made by the peaks in the NOESY with other spin systems.

One important problem using homonuclear nuclear magnetic resonance is overlap between peaks. This occurs when different protons have the same or very similar chemical shifts. This problem becomes greater as the protein becomes larger, so homonuclear nuclear magnetic resonance is usually restricted to small proteins or peptides.

Nitrogen-15 Nuclear Magnetic Resonance

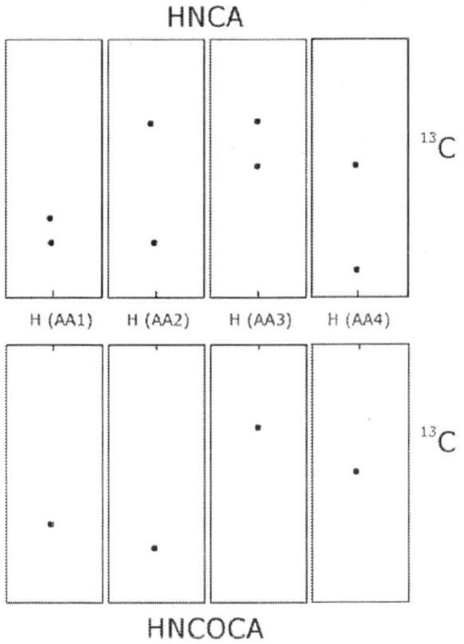

Figure: Schematic of an HNCA and HNCOCA for four sequential residues. The nitrogen-15 dimension is perpendicular to the screen. Each window is focused on the nitrogen chemical shift of that amino acid. The sequential assignment is made by matching the alpha carbon chemical shifts. In the HNCA each residue sees the alpha carbon of itself and the preceding residue. The HNCOCA only sees the alpha carbon of the preceding residue.

Carbon-13 and Nitrogen-15 Nuclear Magnetic Resonance

When the protein is labelled with carbon-13 and nitrogen-15 it is possible to record experiments that transfers magnetisation over the peptide bond, and thus connect different spin systems through bonds. This is usually done using some of the following experiments, HNCO, HN(CA)CO, HNCA, HN(CO)CA, HNCACB and CBCA(CO)NH. All six experiments consist of a ^{1}H-^{15}N plane (similar to a HSQC spectrum) expanded with a carbon dimension. In the HN(CA)CO, each H_N plane contains the peaks from the carbonyl carbon from its residue as well the preceding one in the sequence. The HNCO contains the carbonyl carbon chemical shift from only the preceding residue, but is much more sensitive than HN(CA)CO.

These experiments allow each ^{1}H-^{15}N peak to be linked to the preceding carbonyl carbon, and sequential assignment can then be undertaken by matching the shifts of each spin system's own and previous carbons. The HNCA and HN(CO)CA works similarly, just with the alpha carbons (C_α) rather than the carbonyls, and the HNCACB and the CBCA(CO)NH contains both the alpha carbon and the beta carbon (C_β). Usually several of these experiments are required to resolve overlap in the carbon dimension. This procedure is usually less ambiguous than the NOESY based method, since it is based on through bond transfer. In the NOESY-based methods additional peaks that are close in space but not belonging to the sequential residues will appear confusing the assignment process. When the sequential assignment has been done it is usually possible to extend the assignment from the C_α and C_β to the rest of the sidechain using experiments such as HCCH-TOCSY, which is basically a TOCSY experiment resolved in an additional carbon dimension.

Restraint Generation

In order to make structure calculations a number of experimentally determined restraints have to be generated. These fall into different categories, the most widely used is distance restraints and angle restraints.

Distance Restraints

A crosspeak in a NOESY experiment signifies spatial proximity between the two nuclei in question. Thus each peak can be converted in to a maximum distance between the nuclei, usually between 1.8 and 6 angstroms. The intensity of a noesy peak is proportional to the distance to the minus 6th power, so the distance is determined according to intensity of the peak. The intensity-distance relationship is not exact, so usually a distance range is used.

It is of great importance to assign the noesy peaks to the correct nuclei based on the chemical shifts. If this task is performed manually it is usually very labour intensive, since proteins usually have thousands of noesy peaks. Some computer programs such as UNIO, CYANA and ARIA/CNS perform this task automatically on manually pre-processed listings of peak positions and peak volumes, coupled to a structure calculation. Direct access to the raw NOESY data without the cumbersome need of iteratively refined peak lists is so far only granted by the ATNOS/CANDID approach implemented in the UNIO software package and thus indeed guarantees objective and efficient NOESY spectral analysis.

To obtain as accurate assignments as possible it is a great advantage to have access to carbon-13 and nitrogen-15 noesy experiments, since they help to resolve overlap in the proton dimension. This leads to faster and more reliable assignments, and in turn to better structures.

Angle Restraints

In addition to distance restraints, restraints on the torsion angles of the chemical bonds, typically the psi and phi angles can be generated. One approach is to use the Karplus equation, to generate angle restraints from coupling constants. Another approach uses the chemical shifts to generate angle restraints. Both methods use the fact that the geometry around the alpha carbon affects the coupling constants and chemical shifts, so given the coupling constants or the chemical shifts, a qualified guess can be made about the torsion angles.

Orientation Restraints

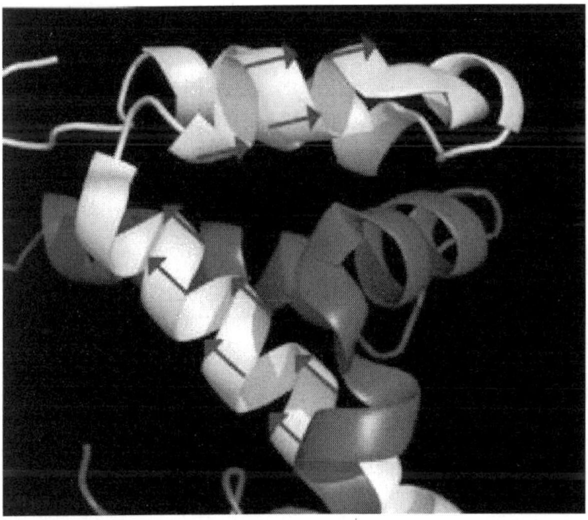

Figure: The blue arrows represent the orientation of the N – H bond of selected peptide bonds. By determining the orientation of a sufficient amount of bonds relative to the external magnetic field, the structure of the protein can be determined. From PDB record 1KBH

The analyte molecules in a sample can be partially ordered with respect to the external magnetic field of the spectrometre by manipulating the sample conditions. Common techniques include addition of bacteriophages or bicelles to the sample, or preparation of the sample in a stretched polyacrylamide gel. This creates a local environment that favours certain orientations of nonspherical molecules. Normally in solution NMR the dipolar couplings between nuclei are averaged out because of the fast tumbling of the molecule. The slight overpopulation of one orientation means that a residual dipolar coupling remains to be observed. The dipolar coupling is commonly used in solid state NMR and provides information about the relative orientation of the bond vectors relativc to a single global reference frame. Typically the orientation of the N-H vector is probed in a HSQC like experiment. Initially residual dipolar couplings were used for refinement of previously determined structures, but attempts at de novo structure determination have also been made.

Hydrogen–Deuterium Exchange

NMR spectroscopy is nucleus specific. Thus it can distinguish between hydrogen and deuterium. The amide protons in the protein exchange readily with the solvent, and if the solvent contains a different isotope, typically deuterium, the reaction can be monitored by NMR spectroscopy. How rapidly a given amide exchanges reflects its solvent accessibility. Thus amide exchange rates can give information on which parts of the protein are buried, hydrogen bonded etc. A common application is to compare the exchange of a free form versus a complex. The amides that become protected in the complex, are assumed to be in the interaction interface.

Hydrogen–deuterium exchange (also called H–D or H/D exchange) is a chemical reaction in which a covalently bonded hydrogen atom is replaced by a deuterium atom, or vice versa. Usually the examined protons are the amides in the backbone of a protein. The method gives information about the solvent accessibility of various parts of thc molecule, and thus the tertiary structure of the protein. Hydrogen exchange was first shown and explored by Kaj Ulrik Linderstrøm-Lang.

The Exchange Reaction

In solution, Amide hydrogens in the peptide bonds of proteins exchange protons with the solvent. By changing the solvent from H_2O to D_2O, deuterons will be incorporated in the amide positions and the exchange reaction can be followed. Most often, deuterium is added to a protein in H_2O by diluting the H_2O solution with D_2O (e.g. tenfold).

Usually exchange is performed at physiological pH (7.0–8.0) where proteins are in their most native ensemble of conformational states.

Because the exchange reaction can be either acid or base catalysed, it is strongly pH dependent. For the backbone amide hydrogens, the minimum exchange rate occurs at approximately pH 2.6, on average. By performing the exchange at neutral pH and then rapidly changing the pH, the exchange rates of the backbone amide hydrogens can be dramatically slowed, or quenched. The pH at which the reaction is quenched depends on the analysis method. For detection by NMR, the pH may be moved to around 4.0–4.5. For detection by mass spectrometry, the pH is dropped to the minimum of the exchange curve, pH 2.6. In the most basic experiment, the reaction is allowed to take place for a set time before it is quenched.

The deuteration pattern of a quenched protein can be stably maintained in aprotic environments. However, analysis of the deuteration is usually performed in an aqueous solution, which means that exchange will continue at a slow rate even after the reaction is quenched. Reversion of deuterated positions after the quench step is referred to as *back-exchange* and various methods have been devised to correct for this.

Detection of H/D Exchange

H-D exchange was measured originally by the father of hydrogen exchange Kaj Ulrik Linderstrøm-Lang using density gradient tubes. In modern times, H/D exchange has primarily been monitored by the methods: NMR spectroscopy and mass spectrometry. Each of these methods have their advantages and drawbacks.

NMR Spectroscopy

Hydrogen and deuterium nuclei are grossly different in their magnetic properties. Thus it is possible to distinguish between them by NMR spectroscopy. Typically HSQC spectra are recorded at a series of timepoints while the hydrogen is exchanging with the deuterium. Since the HSQC experiment is specific for hydrogen, the signal will decay exponentially as the hydrogen exchanges. It is then possible to fit an exponential function to the data, and obtain the exchange constant. This method gives residue-specific information for all the residues in the protein simultaneously. The major drawback is that it requires a prior assignment of the spectrum for the protein in question. This can be very labour intensive, and usually limits the method to proteins smaller than 25 kDa. Because it takes minutes to hours to record the spectra, it is difficult to obtain information about amides, that exchange in shorter time frames.

Mass Spectrometry

Mass spectrometry has several advantages over NMR with respect to analysis of H/D exchange reactions: Much less material is needed, the concentration of protein can be very low (as low as 0.1 uM), the size limit is much greater, and data can usually be collected and interpreted much more quickly.

The deuterium nucleus is twice as heavy as the hydrogen nucleus because it contains a neutron as well as a proton. Thus a protein that contains some deuterium will be heavier than one that contains all hydrogen. As a protein is increasingly deuterated, the molecular mass increases correspondingly. Detecting the change in the mass of a protein upon deuteration was made possible by modern protein mass spectrometry, first reported in 1991 by Katta and Chait.

The location and relative amount of deuterium exchange along the peptide backbone can be determined roughly by subjecting the protein to proteolysis after the exchange reaction has been quenched. Individual peptides are then analysed for overall deuteration of each peptide fragment. Using this technique the resolution of deuterium exchange is determined by the size of the peptides produced during digestion.

Pepsin, an acid protease, is commonly used for proteolysis, as the quench pH must be maintained during the proteolytic reaction. To minimise the back-exchange, proteolysis and subsequent mass spectrometry analysis must be done as quickly as possible. HPLC separation of the peptic digest is often carried out at low temperature just prior to electrospray mass spectrometry to minimise back-exchange. More recently, UPLC has been used due to its superior separation capabilities.

It was proposed in 1999 that it might be possible to achieve single-residue resolution by using collision-induced dissociation fragmentation of deuterated peptides in conjunction with tandem mass spectrometry. It was soon discovered that CID causes "scrambling" of the deuterium position within the peptides. However, fragmentation produced by electron capture dissociation and electron transfer dissociation proceed with little or no scrambling under the correct experimental conditions.

This suggests that eventually it may be possible to obtain single-residue resolution of H/D exchange reactions on a routine basis.

Applications to Protein Structure

It is not possible to determine the structure of a protein with H/D exchange nor is it possible to define secondary structural elements. The reasons for this are related to the way in which protein structure slows exchange. Exchange rates are a function of two parameters: solvent accessibility and hydrogen bonding. Thus an amide which is part of an intramolecular hydrogen bond will exchange slowly if at all, while an amide on the surface of protein hydrogen bonded to water will exchange rapidly. Amides buried from the solvent but not hydrogen bonded may also have very slow exchange rates. Because both solvent accessibility and hydrogen bonding contribute to the rate of exchange, it becomes difficult to attribute a given exchange rate to a structural element without x-ray crystallography or NMR structural data. H/D exchange has been used to characterise the folding pathway of proteins, by refolding the protein under exchange conditions. The parts of the structure that form rapidly, will be protected quickly, and thus not exchanged, whereas areas that fold late in the pathway will be exposed to the exchange for longer periods of time. Thus H/D exchange can be used to determine the sequence of various folding events. The critical factor determining the time resolution of this approach is the time required for quenching.

H/D exchange has been used to characterise protein-protein interactions. The exchange reaction needs to be carried out with the isolated proteins and with the complex. The exchanging regions are then compared. If a region is buried by the binding, the amides in this region may be protected in the complex and exchange slowly. However, one must bear in mind that H-D exchange cannot be used to locate binding interfaces for all protein-protein interactions. Some protein-protein interactions are driven by electrostatic forces of side chains and are unlikely to change the exchange rate of backbone amide hydrogens, particularly if the amide hydrogens are located in stable structural elements such as alpha helicies.

Lastly, H/D exchange can be used to monitor conformational changes in proteins as they relate to protein function. If conformation is altered as result of post-translational modification, enzyme activation, drug binding or other functional events, there will likely be a change to H/D exchange that can be detected.

Structure Calculation

The experimentally determined restraints can be used as input for the structure calculation process. Researchers, using computer programs such as CYANA (Software) or XPLOR-NIH, attempt to satisfy as many of the restraints as possible, in addition to general properties of proteins such as bond lengths and angles. The algorithms convert the restraints and the general protein properties into energy terms, and thus try to minimise the energy. The process results in an ensemble of structures that, if the data were sufficient to dictate a certain fold, will converge.

Dynamics

In addition to structures, nuclear magnetic resonance can yield information on the dynamics of various parts of the protein. This usually involves measuring relaxation times such as T_1 and T_2 to determine order parameters, correlation times, and chemical exchange rates. NMR relaxation is a consequence of local fluctuating magnetic fields within a molecule. Local fluctuating magnetic fields are generated by molecular motions. In this way measurements of relaxation times can provide information of motions within a molecule on the atomic level. In NMR studies of protein dynamics the nitrogen-15 isotope is the preferred nucleus to study because its relaxation times are relatively simple to relate to molecular motions This however requires isotope labelling of the protein. The T_1 and T_2 relaxation times can be measured using various types of HSQC based experiments. The types of motions which can be detected are motions that occur on a time-scale ranging from about 10 picoseconds to about 10 nanoseconds. In addition slower motions, which take place on a time-scale ranging from about 10 microseconds to 100 milliseconds, can also be studied. However, since nitrogen atoms are mainly found in the backbone of a protein, the results mainly reflect the motions of the backbone, which is the most rigid part of a protein molecule. Thus, the results obtained from nitrogen-15 relaxation measurements may not be representative for the whole protein. Therefore techniques utilising relaxation measurements of carbon-13 and deuterium have recently been developed, which enables systematic studies of motions of the amino acid side chains in proteins.

NMR Spectroscopy on Large Proteins

Traditionally nuclear magnetic resonance spectroscopy has been limited to relatively small proteins or protein domains. This is in part caused by problems resolving overlapping peaks in larger proteins, but this has been alleviated by the introduction of isotope labelling and multidimensional experiments. Another more serious problem is the fact that in large proteins the magnetization relaxes faster, which means there is less time to detect the signal. This in turn causes the peaks to become broader and weaker, and eventually disappear. Two techniques have been introduced to attenuate the relaxation: transverse relaxation optimized spectroscopy (TROSY) and deuteration of proteins. By using these techniques it has been possible to study proteins in complex with the 900 kDa chaperone GroES-GroEL.

Automation of the Process

Structure determination by NMR has traditionally been a time consuming process, requiring interactive analysis of the data by a highly trained scientist. There has been a considerable interest in automating the process to increase the throughput of structure determination and to make protein NMR accessible to non-experts. The two most time consuming processes involved are the sequence-specific resonance assignment (backbone and side-chain assignment) and the NOE assignment tasks. Several different computer programs have been published that target individual parts of the overall NMR structure determination process in an automated fashion. Most progress has been achieved for the task of automated NOE assignment. So far, only the FLYA and the UNIO approach were proposed to perform the entire protein NMR structure determination process in an automated manner without any human intervention. Efforts have also been made to standardize the structure calculation protocol to make it quicker and more amenable to automation.

Nucleic Acids

"Nucleic acid NMR" is the use of NMR spectroscopy to obtain information about the structure and dynamics of polynucleic acids, such as DNA or RNA. As of 2003, nearly half of all known RNA structures had been determined by NMR spectroscopy.

Nucleic acid and protein NMR spectroscopy are similar but differences exist. Nucleic acids have a smaller percentage of hydrogen atoms, which are the atoms usually observed in NMR spectroscopy, and because nucleic acid double helices are stiff and roughly linear, they do not fold back on themselves to give "long-range" correlations.

The types of NMR usually done with nucleic acids are ^1H or proton NMR, ^{13}C NMR, ^{15}N NMR, and ^{31}P NMR. Two-dimensional NMR methods are almost always used, such as correlation spectroscopy (COSY) and total coherence transfer spectroscopy (TOCSY) to detect through-bond nuclear couplings, and nuclear Overhauser effect spectroscopy (NOESY) to detect couplings between nuclei that are close to each other in space.

Parameters taken from the spectrum, mainly NOESY cross-peaks and coupling constants, can be used to determine local structural features such as glycosidic bond angles, dihedral angles (using the Karplus equation), and sugar pucker conformations. For large-scale structure, these local parameters must be supplemented with other structural assumptions or models, because errors add up as the double helix is traversed, and unlike with proteins, the double helix does not have a compact interior and does not fold back upon itself. NMR is also useful for investigating nonstandard geometries such as bent helices, non-Watson–Crick basepairing, and coaxial stacking. It has been especially useful in probing the structure of natural RNA oligonucleotides, which tend to adopt complex conformations such as stem-loops and pseudoknots. NMR is also useful for probing the binding of nucleic acid molecules to other molecules, such as proteins or drugs, by seeing which resonances are shifted upon binding of the other molecule.

Nucleic acid NMR is the use of nuclear magnetic resonance spectroscopy to obtain information about the structure and dynamics of nucleic acid molecules, such as DNA or RNA. It is useful for molecules of up to 100 nucleotides, and as of 2003, nearly half of all known RNA structures had been determined by NMR spectroscopy.

NMR has advantages over X-ray crystallography, which is the other method for high-resolution nucleic acid structure determination, in that the molecules are being observed in their natural solution state rather than in a crystal lattice that may affect the molecule's structural properties. It is also possible to investigate dynamics with NMR. This comes at the cost of slightly less accurate and detailed structures than crystallography.

Nucleic acid NMR uses techniques similar to those of protein NMR, but has several differences. Nucleic acids have a smaller percentage of hydrogen atoms, which are the atoms usually observed in NMR, and because nucleic acid double helices are stiff and roughly linear, they do not fold back on themselves to give "long-range" correlations. Nucleic acids also tend to have resonances distributed over a smaller range than proteins, making the spectra potentially more crowded and difficult to interpret.

Experimental Methods

Two-dimensional NMR methods are almost always used with nucleic acids. These include correlation spectroscopy (COSY) and total coherence transfer spectroscopy (TOCSY) to detect through-bond nuclear couplings, and nuclear Overhauser effect spectroscopy (NOESY) to detect couplings between nuclei that are close to each other in space. The types of NMR usually done with nucleic acids are ^{1}H NMR, ^{13}C NMR, ^{15}N NMR, and ^{31}P NMR. ^{19}F NMR is also useful if nonnatural nucleotides such as 2'-fluoro-2'-deoxyadenosine are incorporated into the nucleic acid strand, as natural nucleic acids do not contain any fluorine atoms.

^{1}H and ^{31}P have near 100% natural abundance, while ^{13}C and ^{15}N have low natural abundances.

For these latter two nuclei, there is the capability of isotopically enriching desired atoms within the molecules, either uniformly or in a site-specific manner. Nucleotides uniformly enriched in ^{13}C and/or ^{15}N can be obtained through biochemical methods, by performing polymerase chain reaction using dNTPs or NTPs derived from bacteria grown in an isotopically enriched environment. Site-specific isotope enrichment must be done through chemical synthesis of the labelled nucleoside phosphoramidite monomer and of the full strand; however these are difficult and expensive to synthesize.

Because nucleic acids have a relatively large number of protons which are solvent-exchangable, nucleic acid NMR is generally not done in D_2O solvent as is common with other types of NMR. This is because the deuterium in the solvent would replace the exchangeable protons and extinguish their signal. H_2O is used as a solvent, and other methods are used to eliminate the strong solvent signal, such as saturating the solvent signal before the normal pulse sequence ("presaturation"), which works best a low temperature to prevent exchange of the saturated solvent protons with the nucleic acid protons; or exciting only resonances of interest ("selective excitation"), which has the additional, potentially undesired effect of distorting the peak amplitudes.

Structure Determination

The exchangeable and non-exchangable protons are usually assigned to their specific peaks as two independent groups. For exchangeable protons, which are for the most part the protons involved in base pairing, NOESY can be used to find through-space correlations between on neighbouring bases, allowing an entire duplex molecule to be assigned through sequential walking. For nonexchangable protons, many of which are on the sugar moiety of the nucleic acid, COSY and TOCSY are used to identify systems of coupled nuclei, while NOESY is again used to correlate the sugar to the base and each base to its neighbouring base.

For duplex DNA nonexchangable protons the H6/H8 protons on the base correlate to their counterparts on neighbouring bases and to the H1' proton on the sugar, allowing sequential walking to be done. For RNA, the differences in chemical structure and helix geometry make this assignment more technically difficult, but still possible. The sequential walking methodology is not possible for non-double helical nucleic acid structures, nor for the Z-DNA form, making assignment of resonances more difficult.

Parameters taken from the spectrum, mainly NOESY cross-peaks and coupling constants, can be used to determine local structural features such as glycosidic bond angles, dihedral angles (using the Karplus equation), and sugar pucker conformations.

The presence or absence of inimo proton resonances, or of coupling between ^{15}N atoms across a hydrogen bond, indicates the presence or absence of basepairing. For large-scale structure, these local parameters must be supplemented with other structural assumptions or models, because errors add up as the double helix is traversed, and unlike with proteins, the double helix does not have a compact interior and does not fold back upon itself. However, long-range orientation information can be obtained through residual dipolar coupling experiments in a medium which imposes a weak alignment on the nucleic acid molecules.

NMR is also useful for investigating nonstandard geometries such as bent helices, non-Watson–Crick basepairing, and coaxial stacking. It has been especially useful in probing the structure of natural RNA oligonucleotides, which tend to adopt complex conformations such as stem-loops and pseudoknots. Interactions between RNA and metal ions can be probed by a number of methods, including observing changes in chemical shift upon ion binding, observing line broadening for paramagnetic ion species, and observing intermolecular NOE contacts for organometallic mimics of the metal ions. NMR is also useful for probing the binding of nucleic acid molecules to other molecules, such as proteins or drugs. This can be done by chemical-shift mapping, which is seeing which resonances are shifted upon binding of the other molecule, or by cross-saturation experiments where one of the binding molecules is selectively saturated and, if bound, the saturation transfers to the other molecule in the complex.

Dynamic properties such as duplex–single strand equilibria and binding rates of other molecules to duplexes can also be determined by its effect on the spin–lattice relaxation time T_1, but these methods are insensitive to intermediate rates of 10^4–10^8 s^{-1}, which must be investigated with other methods such as solid-state NMR. Dynamics of mechanical properties of a nucleic acid double helix such as bending and twisting can also be studied using NMR. Pulsed field gradient NMR experiments can be used to measure diffusion constants.

Chapter 9

Photochemistry

Photochemistry, a sub-discipline of chemistry, is the study of chemical reactions that proceed with the absorption of light by atoms or molecules. Everyday examples include photosynthesis, the degradation of plastics and the formation of vitamin D with sunlight.

Principles

Light is a type of electromagnetic radiation, a source of energy. The Grotthuss–Draper law (for chemists Theodor Grotthuss and John W. Draper), states that light must be absorbed by a chemical substance in order for a photochemical reaction to take place. For each photon of light absorbed by a chemical system, no more than one molecule is activated for a photochemical reaction, as defined by the quantum yield.

Chemical reactions occur only when a molecule is provided the necessary "activation energy". A simple example can be the combustion of gasoline (a hydrocarbon) into carbon dioxide and water. In this reaction, the activation energy is provided in the form of heat or a spark. In case of photochemical reactions light provides the activation energy. Simplistically, light is one mechanism for providing the activation energy required for many reactions. If laser light is employed, it is possible to selectively excite a molecule so as to produce a desired electronic and vibrational state. Equally, the emission from a particular state may be selectively monitored, providing a measure of the population of that state.

If the chemical system is at low pressure, this enables scientists to observe the energy distribution of the products of a chemical reaction before the differences in energy have been smeared out and averaged by repeated collisions. The absorption of a photon of light by a reactant molecule may also permit a reaction to occur not just by bringing the molecule to the necessary activation energy, but also by changing the symmetry of the molecule's electronic configuration, enabling an otherwise inaccessible reaction path, as described by the Woodward-Hoffmann selection rules.

A 2+2 cycloaddition reaction is one example of a pericyclic reaction that can be analysed using these rules or by the related frontier molecular orbital theory. Photochemical reactions involve electronic reorganisation initiated by electromagnetic radiation. The reactions are several orders of magnitude faster than thermal reactions; reactions as fast as 10^{-9} seconds and associated processes as fast as 10^{-15} seconds are often observed.

Spectral Regions

Photochemists typically work in only a few sections of the electromagnetic spectrum. Some of the most widely used sections, and their wavelengths, are the following:

- Ultraviolet: 100–400 nm
- Visible Light: 400–700 nm
- Near infrared: 700–2500 nm

Applications

Many important processes involve photochemistry. The premier example is photosynthesis, in which most plants use solar energy to convert carbon dioxide and water into glucose, disposing of oxygen as a side-product. Humans rely on photochemistry for the formation of vitamin D.

In fireflies, an enzyme in the abdomen catalyses a reaction that results in bioluminescence. Photochemistry can also be highly destructive. Medicine bottles are often made with darkened glass to prevent the drugs from photodegradation. A pervasive reaction is the generation of singlet oxygen by photosensitized reactions of triplet oxygen.

Typical photosensitizers include tetraphenylporphyrin and methylene blue. The resulting singlet oxygen is an aggressive oxidant, capable of converting C-H bonds into C-OH groups. In photodynamic therapy, light is used to destroy tumours by the action of singlet oxygen. Many polymerizations are started by photoinitiatiors, which decompose upon absorbing light to produce the free radicals for Radical polymerization.

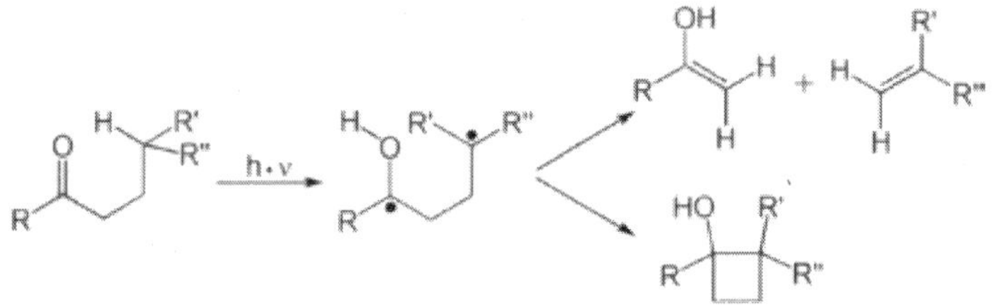

Figure:

In the area of photochemistry, a photochemical reaction is a chemical reaction that is induced by light. Photochemical reactions are valuable in organic and inorganic chemistry because they proceed differently than thermal reactions. Photochemical reactions are not only very useful but also can be a serious nuisance, as in the photodegradation of many materials, e.g. polyvinyl chloride. A large-scale application of photochemistry is photoresist technology, used in the production of microelectronic components. Vision is initiated by a photochemical reaction of rhodopsin.

Experimental Set-up

Photochemical reactions require a light source that emits wavelengths corresponding to an electronic transition in the reactant. In the early experiments (and in everyday life), sunlight was the light source, although it is polychromatic. Mercury-vapour lamps are more common in the laboratory. Low pressure mercury vapour lamps mainly emit at 254 nm. For polychromatic sources, wavelength ranges can be selected using filters. Alternatively, LEDs and Rayonet lamps emit monochromatically.

The emitted light must of course reach the targeted functional group without being blocked by the reactor, medium, or other functional groups present. For many applications, quartz is used for the reactors as well as to contain the lamp.

Pyrex absorbs at wavelengths shorter than 275 nm. The solvent is an important experimental parameter. Solvents are potential reactants and for this reason, chlorinated solvents are avoided because the C-Cl bond can lead to chlorination of the substrate. Strongly absorbing solvents prevent photons from reaching the substrate. Hydrocarbon solvents absorb only at short wavelengths and are thus preferred for photochemical experiments requiring high energy photons. Solvents containing unsaturation absorb at longer wavelengths and can usefully filter out short wavelengths. For example, cyclohexane and acetone "cut off" (absorb strongly) at wavelengths shorter than 215 and 330 nm, respectively.

Excitation

Photoexcitation is the first step in a photochemical process where the reactant is elevated to a state of higher energy, an excited state. The photon can be absorbed directly by the reactant or by a photosensitizers, which absorbs the photon and transfers the energy to the reactant. The opposite process is called quenching when a photoexited state is deactivated by a chemical reagent. Most photochemical transformations occur through a series of simple steps known as primary photochemical processes. One common example of these processes is the excited state proton transfer (ESPT).

Organic Photochemistry

Examples of photochemical organic reactions are electrocyclic reactions, photoisomerization and Norrish reactions. Alkenes undergo many important reactions that proceed via a photon-induced π to π^* transition. The first electronic excited state of an alkene lack the π-bond, so that rotation about the C-C bond is rapid and the molecule engages in reactions not observed thermally. These reactions include cis-trans isomerization, cycloaddition to other (ground state) alkene to give cyclobutane derivatives. The cis-trans isomerization of a (poly)alkene is involved in retinal, a component of the machinery of vision. The dimerization of alkenes is relevant to the photodamage of DNA, where thymine dimers are observed upon illuminating DNA to UV radiation. Such dimers interfere with transcription. The beneficial effects of sunlight are associated with the photochemically induced retro-cyclization (decyclization) reaction of ergosterol to give vitamin D. In the DeMayo reaction, an alkene reacts with a 1,3-diketone reacts via its enol to yield a 1,5-diketone. Still another common photochemical reaction is Zimmerman's Di-pi-methane rearrangement.

In an industrial application, about 100,000 tonnes of benzyl chloride are prepared annually by the gas-phase photochemical reaction of toluene with chlorine. The light is absorbed by chlorine molecule, the low energy of this transition being indicted by the yellowish colour of the gas. The photon induces homolysis of the Cl-Cl bond, and the resulting chlorine radical converts toluene to the benzyl radical:

$$Cl_2 + h\nu \rightarrow 2\ Cl\cdot$$

$C_6H_5CH_3 + Cl\cdot \rightarrow C_6H_5CH_2\cdot + HCl$

$C_6H_5CH_2\cdot + Cl\cdot \rightarrow C_6H_5CH_2Cl$

Mercaptans can be produced by photochemical addition of hydrogen sulfide (H_2S) to alpha olefins.

Inorganic and Organometallic Photochemistry

Coordination complexes and organometallic compounds are also photoreactive. These reactions can entail cis-trans isomerization. More commonly photoreactions result in dissociation of ligands, since the photon excites an electron on the metal to an orbital that is antibonding with respect to the ligands. Thus, metal carbonyls that resist thermal substitution undergo decarbonylation upon irradiation with UV light. UV-irradiation of a THF solution of molybdenum hexacarbonyl gives the THF complex, which is synthetically useful:

$Mo(CO)_6 + THF \rightarrow Mo(CO)_5(THF) + CO$

In a related reaction, photolysis of iron pentacarbonyl affords diiron nonacarbonyl:

$2\ Fe(CO)_5 \rightarrow Fe_2(CO)_9 + CO$

Carbon Nanotubes

Certain species of carbon nanotubes also undergo photochemical reactions when exposed to intense pulsed light from a laser, arc lamp, or flashbulb. In the presence of air, Carbon Nanotubes will undergo combustion at high temperature (700-1500C). Under inert atmospheres, the nanotubes will break open and reorganise into larger "nanohorn" structures. Although bleaching has long been practiced, the first photochemical reaction was described by Trommsdorf in 1834. He observed that crystals of the compound α-santonin when exposed to sunlight turned yellow and burst. In a 2007 study the reaction was described as a succession of three steps taking place within a single crystal. The first step is a rearrangement reaction to a cyclopentadienone intermediate 2, the second one a dimerization in a Diels-Alder reaction (3) and the third one a intramolecular [2+2]cycloaddition (4). The bursting effect is attributed to a large change in crystal volume on dimerization.

Mechanistic Organic Photochemistry

Mechanistic organic photochemistry is that aspect of organic photochemistry which seeks to explain the mechanisms of organic photochemical reactions. The absorption of ultraviolet light by organic molecules very often leads to reactions. In the earliest days sunlight was employed while in more modern times ultraviolet lamps are employed.

Organic photochemistry has proven to be a very useful synthetic tool. Complex organic products can be obtained simply. Over the last century and earlier an immense number of photochemical reactions have been uncovered.

In modern times the field is quite well understood and is used in organic synthesis and industrially. The utility of organic photochemistry has arisen only by virtue of the available mechanistic treatment; reactions which appear unlikely in ground-state understanding become understandable and accessible in terms of electronic excited-state consideration.

One of the earliest photochemical studies dealt with the natural product santonin. In the 19th century it had been observed by Ciamician that in Italian sunlight santonin gave several photoproducts. The structure of santonin was first correctly described by Clemo and Hayworth in 1929. The initial photoproduct obtained from santonin is lumisantonin. As depicted in Eqn. 1, the photoreaction involves a rearrangement. Using steroid numbering, we note that the C-3 carbonyl group has moved to C-2, the C-4 methyl has moved to C-1, and the C-10 carbon has been inverted.

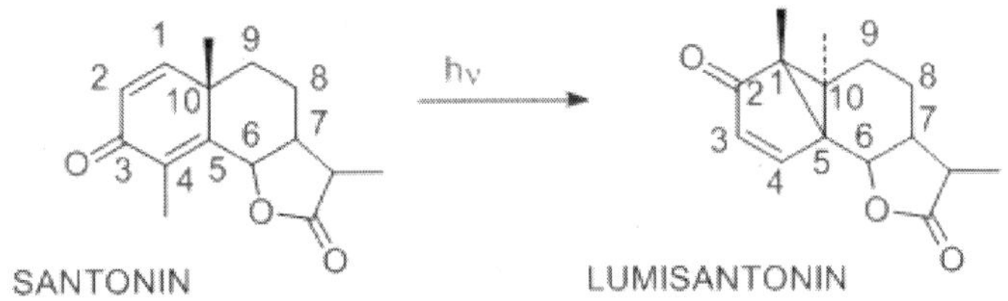

Figure: An comparably bizarre example was uncovered by Egbert Havinga in 1956. The curious result was activation on photolysis by a meta nitro group in contrast to the usual activation by ortho and para groups.

Over the decades, many interesting but puzzling organic photochemical reactions were discovered that did not proceed by ordinary organic ground state processes. Rather, they arose from the electronic excited states. The real problem was that, at the time, organic chemists were not versed in quantum mechanics and physical chemists were not versed in organic chemistry. Real mechanistic treatments were not possible.

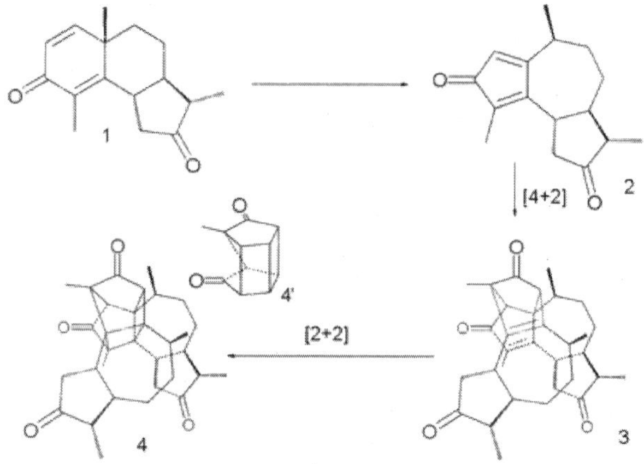

Figure:

Starting in 1961 it was found that one could understand organic photochemical reactions in the context of the relevant excited states. One example is the n-pi* excitation of mono-carbonyl compounds, the simplest being that of formaldehyde. The structure was first described by Mulliken. The three-dimensional representation (top drawing) is simplified in the second line using a two-dimensional representation, which facilitates arrow pushing.

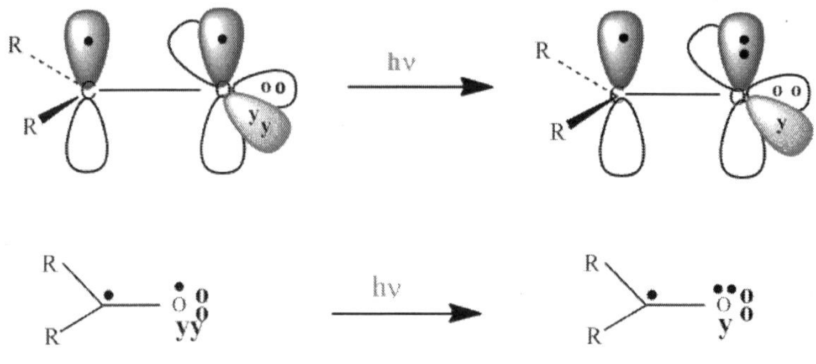

Solid dots represent pi-system electrons; 0 dots represent sp-hybrids; y's are p_y electrons

Figure:

In this early research simple Hückel computations were used to get excited state electron densities and bond-orders. The stereochemistry in Scheme 1 is shown three-dimensionally. The Hückel computations revealed that the beta-carbons (i.e. C2 and C5) of the cyclohexadienone ring had a large bond-order. As seen in the scheme a beta-beta bond is formed. Subsequent to this, radiationless decay leads to a zwitterion ground state. The final rearrangement leads to lumisantonin as can be discerned by comparing the three-dimensional drawing with the earlier two-dimensional representation.

Scheme 1. The santonin to lumisantonin rearrangement in three-dimensions

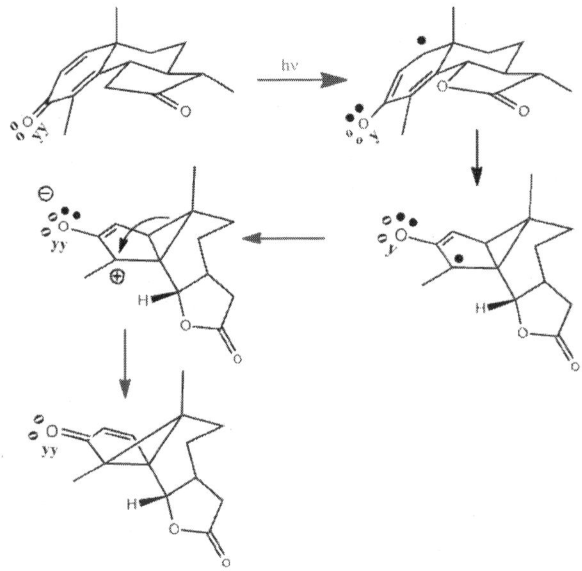

Figure:

4,4-Diphenylcyclohexadienone Rearrangement

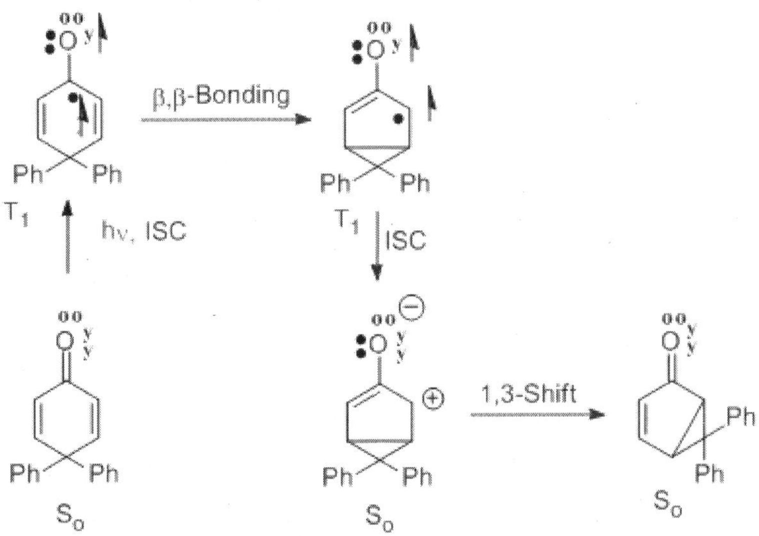

Figure:

Quite parallel to the santonin to lumisantonin example is the rearrangement of 4,4-diphenylcyclohexadienone. Here the n-pi* triplet excited state undergoes the same beta-beta bonding. This is followed by intersystem crossing (i.e. ISC) to form the singlet ground state which is seen to be a zwitterion. The final step is the rearrangement to the bicyclic photoproduct. The reaction is termed the type A cyclohexadienone rearrangement.

The Contrasting Case of 4,4-Diphenylcyclohexenone, One Double Bond Missing

To provide further evidence on the mechanism of the dienone in which there is bonding between the two double bonds, the case of 4,4-diphenylcyclohexenone is presented here. It is seen that the rearrangement is quite different; thus two double bonds are required for a type A rearrangement. With one double bond one of the phenyl groups, originally at C-4, has migrated to C-3 (i.e. the beta carbon).

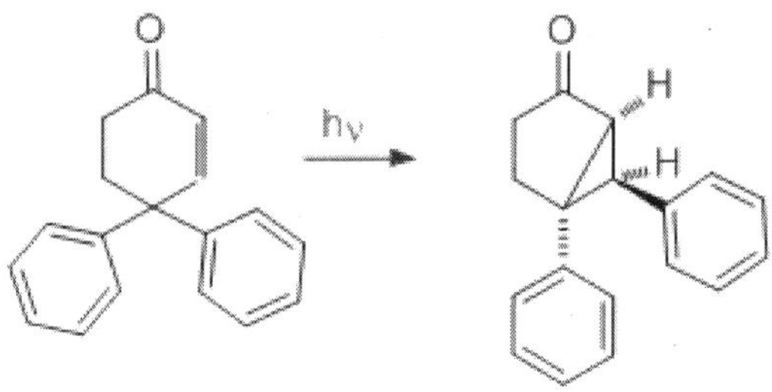

Figure:

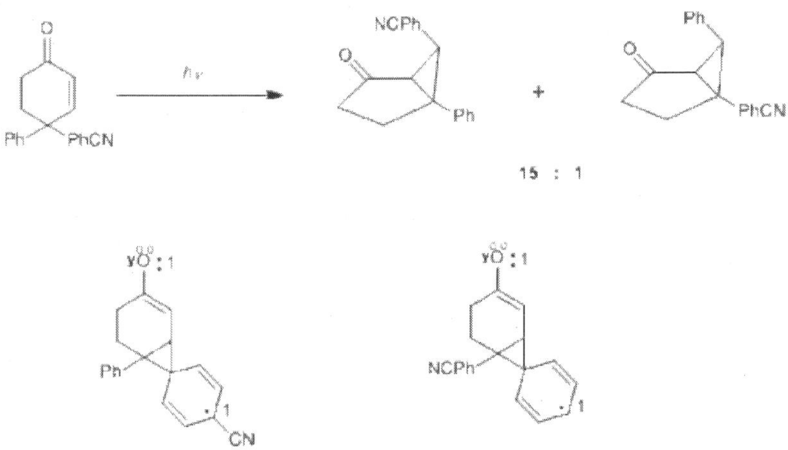

Figure:

It is of considerable interest that when one of the aryl groups has a para-cyano or para-methoxy group, that substituted aryl group migrates in preference. Inspection of the alternative phenonium-type species, in which an aryl group has begun to migrate to the beta-carbon, reveals the greater electron delocalization with a substituent para on the migrating aryl group and thus a more stabilised pathway.

Pi-pi Reactivity*

Still another type of photochemical reaction is the di-pi-methane rearrangement. Two further early examples were the rearrangement of 1,1,5,5-tetraphenyl-3,3-dimethyl-1,4-pentadiene (the "Mariano" molecule) and the rearrangement of barrelene to semibullvalene. We note that, in contrast to the cyclohexadienone reactions which used n-pi* excited states, the di-pi-methane rearrangements utilise pi-pi* excited states.

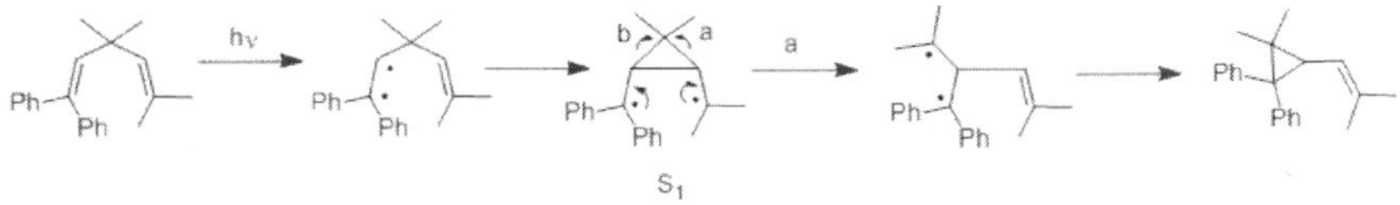

Figure:

Parallel Studies on Multiplicity; the Role of Triplets

Parallel to the structural studies of the Zimmerman group as described above, the Cal. Tech. group with George Hammond pursued the role of multiplicity on reactivity. The importance of triplet excited species was emphasized. The triplets tend to be longer lived than singlets and of lower energy than the singlet of the same configuration. Hammond had summarised the energies of triplets of the more common molecules. He noted that triplets may arise from (A) conversion of the initially formed singlets or by (B) interaction with a higher energy triplet (sensitization). Another contribution of Hammond was the determination of triplet reaction rates. Finally, with low energy triplets present it was shown possible to quench a triplet reaction. The ground work had been laid by very earlier studies by Terenin and Ermolaev in Russia who demonstrated intermolecular triplet transfer at low temperatures and described the kinetics.

Further on the Subsequent Years

After the basics had been established organic photochemistry seemed to accelerate exponentially. One facet derived from the Cal.

Tech. group of George Hammond, a group of remarkable researchers. Nicholas (Nick) Turro was a graduate student in the Hammond group along with Angelo Lamola, Peter Leermakers, Jack Saltiel, Robert Liu, Angelo Lamola and a number of others. However, the emphasis of the Cal. Tech. group remained on observation of kinetic and multiplicity dependence rather than the effect of the electronic structure of the excited species in controlling charges, bond-orders and thus reactivity.

Common Organic Photochemical Reactions

Among the more common organic photochemical reactions there are the Norrish Type I, the Norrish Type II, the racemization of optically active biphenyls, the type A cyclohexadienone rearrangement, the type B cyclohexenone rearrangement, the di-pi-methane rearrangement, the type B bicyclo[3.1.0]hexanone rearrangement to phenols, photochemical electrocyclic processes, the rearrangement of epoxyketones to beta-diketones, ring opening of cyclopropyl ketones, heterolysis of 3,5-dimethoxylbenzylic derivatives, and photochemical cyclizations of dienes. Some of these have been described above and with this being an encyclopedic survey only a selected few are considered here.

The Basic Laws of Photochemistry

The First Law of Photochemistry states that light must be absorbed for photochemistry to occur. This is a simple concept, but it is the basis for performing photochemical and photobiological experiments correctly. If light of a particular wavelength is not absorbed by a system, no photochemistry will occur, and no photobiological effects will be observed, no matter how long one irradiates with that wavelength of light.

The Second Law of Photochemistry states that for each photon of light absorbed by a chemical system, only one molecule is activated for a photochemical reaction. This law is true for ordinary light intensities, however, with high-powered lasers, two-photon reactions can occur, i.e., the molecule is raised to a higher energy state than produced by single-photon absorption.

The Bunsen-Roscoe Law of Reciprocity states that a photochemical effect is directly proportional to the total energy dose, irrespective of the time required to deliver the dose. This law is true for chemicals in a test tube, but the response of cells to radiation usually involves a sequence of interacting biological reactions, making a linear "dose x time" relationship highly unlikely. There is no reciprocity when damage is produced, e.g., DNA damage, but there can be reciprocity over a narrow range of doses for photoreceptors that trigger a response, such as phytochrome.

Electromagnetic Radiation

Electromagnetic radiation consists of waves of electric and magnetic fields travelling in space at right angles to one another.

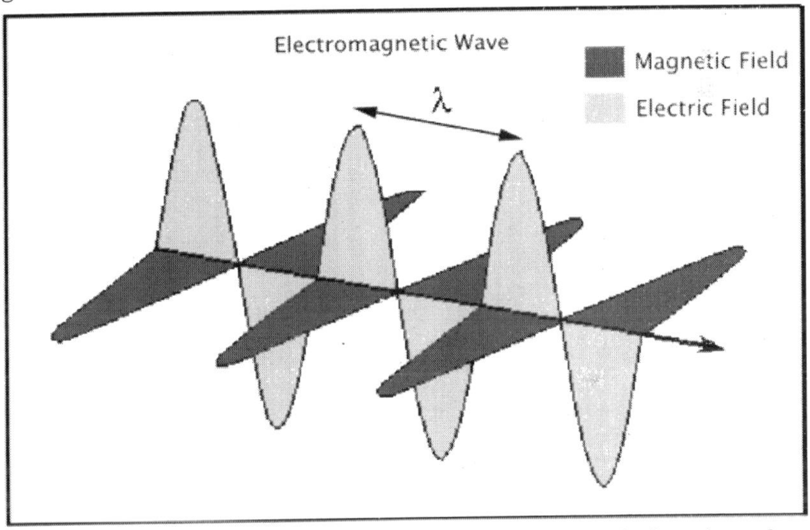

Figure: An electromagnetic wave showing the perpendicularly-oriented waves of electric and magnetic fields, and the characteristic wavelength (λ) of the radiation.

The electromagnetic spectrum is composed of different wavelengths of light having different photon energies, and is classified into the regions. Note that the regions of interest for photochemistry, i.e., visible and ultraviolet (UV), are only a small part of the full electromagnetic spectrum. Longer wavelengths, e.g., far infrared, tend to cause the vibrational excitation of molecules, which results in heating. Shorter wavelength X-rays cause ionization.

The Jablonski Diagram

The energy gained by a molecule when it absorbs a photon causes an electron to be promoted to a higher electronic energy level. The symbols So, S1, T2, etc., refer to the ground electronic state (So), first excited singlet state (S1), second excited triplet state (T2), and so on.

The horizontal lines represent the vibrational levels of each electronic state. Straight arrows indicate radiative transitions, and curly arrows indicate non-radiative transitions. The boxes detail the electronic spins in each orbital, with electrons shown as up and down arrows, to distinguish their spin.

Note that all transitions from one electronic state to another originate from the lowest vibrational level of the initial electronic state. For example, fluorescence occurs only from S1, because the higher singlet states (S2, etc.) decay so rapidly by internal conversion that fluorescence from these states cannot compete.

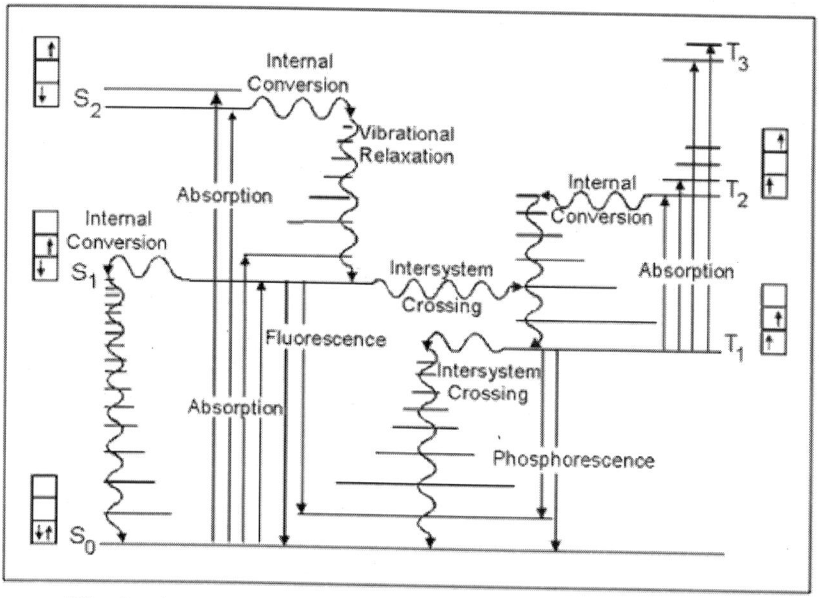

Figure: The basic concepts of this Jablonski diagram are presented in the Basic Photophysics module. This version emphasizes the spins of electrons in each of the singlet states (paired, i.e., opposite orientation, spins) compared to the triplet states (unpaired, i.e., same orientation, spins).

Electronically Excited States

The absorption of a UV or visible photon by a molecule produces an electronically excited state. The distribution of the electrons surrounding the nuclei change, as well as the forces between the atomic nuclei of a molecule. As a result, molecules in electronically excited states often have very different chemical and physical properties than their electronic ground states. For example, α-hydroxy naphthalene becomes a strong acid in its excited state. The ground state of -hydroxy naphthalene has a pKa of 9.2, but this is reduced to 0.4 in the excited singlet state. Molecules such as this are known as photoacids.

The Beer-Lambert Law

The absorption of photons of light is described by the Beer-Lambert Law, a relationship that indicates a decrease in intensity as a beam passes through a medium that can absorb it. Consider a parallel beam of monochromatic light of initial intensity, lo, passing through a homogeneous absorbing medium.

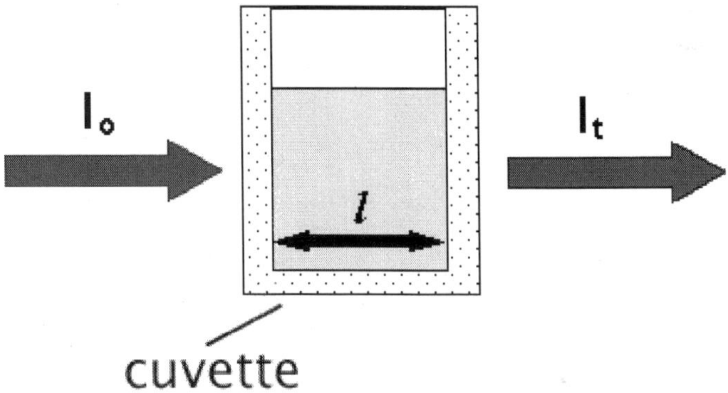

Figure: Schematic representation showing that light of initial intensity, lo, passing through an absorbing medium in a cuvette with light path, l, will emerge with a final intensity, lt. In spectroscopy, absorbance, A, and Optical Density, OD, are used somewhat interchangeably. Optical Density can be expressed as:

$$OD = \log10(1/T)$$

where T is the transmittance. It can also be expressed as:

$$A = OD = -\log10(lt/lo)$$

Note the log scale. An OD of 1 will have a transmittance of 0.1, and a % transmission of 10, but at OD = 2, the transmittance is 0.01, and the % transmission = 1. Another way of expressing this information is to use the Beer-Lambert Law. It states that the absorbance, A, of a molecular species is linearly related to the path length (centimetre), l, the absorber concentration (moles/litre), c, and the proportionality constant, ε, called the molar extinction coefficient of the absorbing molecular species (litres/mole-cm) [a measure of how strongly a chemical species absorbs light at a given wavelength].

$$A = cl$$

Energy Level Diagram

This diagram, known as a *Franck-Condon* energy level diagram, shows potential energy curves for the ground state (So), and first excited singlet state (S1) of an organic molecule as a function of nuclear configuration. These curves are sometimes referred to as potential energy wells, because of their shape. The horizontal lines within each curve represent the vibrational levels of each electronic state. The lowest vibrational state for each energy level is designated as 0, and the levels above it are successively 1, 2, etc. The band assignments in brackets (*e.g.*, (0, 1)) indicate, respectively, the vibrational level of the initial state, and of the final state involved in a transition.

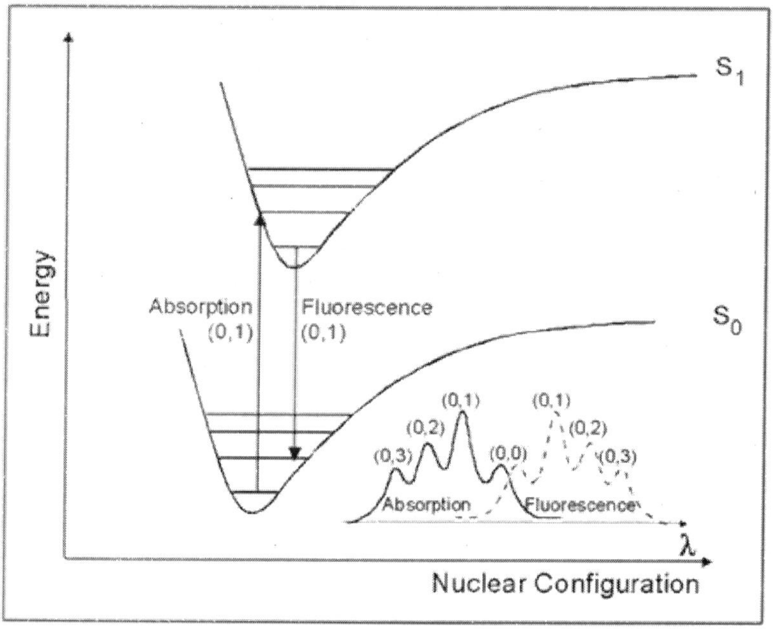

Figure: Franck-Condon Energy Level Diagram

The horizontal axis is the nuclear configuration, which can be thought of as the distance between nuclei. When considering two atoms bonded to each other, the bottom of the well corresponds to the equilibrium bond length. Because excitation involves the movement of charge density into an antibonding orbital, the equilibrium bond length in S1 is generally longer than in S0.

The absorption of light takes place on a much faster time scale (~ 10-15 s) than molecular vibration (~ 10-12 s), and hence the initially formed excited state must have the same nuclear configuration as the ground state.

This transition is called the vertical or Franck-Condon transition, and results in the molecule having excess vibrational energy. The excess vibrational energy can be dissipated through the process of vibrational relaxation, i.e., the process of internal conversion, which returns the molecule to the lowest vibrational level of S1. Fluorescence usually occurs from the lowest vibrational level of S1. Because these transitions occur at lower energies than absorption, fluorescence is observed at longer wavelengths (λ) than absorption (i.e., lower energy).

Quantum Yields and Lifetimes

The energy that a molecule gains when it absorbs light is subsequently lost by a molecule in one of several ways. As shown graphically in the Jablonski diagram, it can lose the energy as heat as it returns to the ground state (internal conversion). Alternately, it can lose the energy as light (fluorescence), usually on a nanosecond time scale. A third pathway is intersystem crossing to a triplet state, from which energy can also be lost as light (phosphorescence), but over much longer times (microseconds or longer). And finally, the energy can be transferred to another molecule.

The *quantum yield* of a process is the probability that an absorbed photon undergoes one particular process. Thus, one can define a quantum yield for fluorescence, a quantum yield for phosphorescence, or a quantum yield for other pathways. Each quantum yield is typically a number between 0 and 1 (except under unusual circumstances), and the total of all quantum yields for a particular absorption event should sum to one. Note that these processes are competing. If conditions are altered such that the quantum yield for fluorescence is increased, then the quantum yield for some other process(es) must decrease.

Consider a molecule, M, that is exposed to light, and absorbs photons at the rate, Iabs. As shown in the following formulas, the excited singlet state of the molecule, 1M*, can fluoresce emitting a photon, $h_{\nu}f$. It can lose energy as heat, and move to the triplet excited state, 3M*, by intersystem crossing (*isc*). Finally, it can lose energy as heat, and move to the ground state by internal conversion (*ic*). The rate of each of these loss processes will be proportional to the concentration of the excited singlet state, 1M*, and a rate constant, *k*, for each process, as given in the second column below.

	Rate
$M \rightarrow {}^1M^*$	I_{abs}
${}^1M^* \xrightarrow{k_f} M + h\nu_f$	$k_f[{}^1M^*]$
${}^1M^* \xrightarrow{k_{isc}} {}^3M^* + heat$	$k_{isc}[{}^1M^*]$
${}^1M^* \xrightarrow{k_{ic}} M + heat$	$k_{ic}[{}^1M^*]$

Because the lifetime of the singlet excited state is relatively short, we can assume that all of the excited singlet states that are formed by light absorption will rapidly decay through one of the three means just described. This is the "steady state" approximation. Since the rate of formation of the excited singlet state is I_{abs}, then the sum of the rates of loss, each given above, must be equal to the rate of formation.

$$I_{abs} = \sum_i K_1 \left[{}^1M^* \right]$$

From this relationship, one can determine the quantum yield for each process. It is simply the rate of that process as a fraction of all pathways for the loss of energy.

Intermolecular Processes: Excited State Quenching

When a second molecule interacts with a molecule in an excited state, new ways may be created for the excited state species to lose its energy of excitation. Such interactions (collisions) can induce the loss of energy in the form of heat, which is called physical quenching, or it can cause the energy to be transferred to the second molecule with or without the transfer of an electron. The former is called simply energy transfer, and the latter electron transfer. Formally, one can write:

$$M^* + Q \rightarrow M + Q + heat \quad physical\,quenching$$

$$M^* + Q \rightarrow M + Q^* \quad energy\,transfer$$

$$M^* + Q \rightarrow M + Q^- \quad energy\,transfer$$

When this kind of quenching occurs it reduces the concentration of the excited state more rapidly than if the quencher were not present. This means that fluorescence, which is proportional to the concentration of the excited state will also be reduced. By monitoring fluorescence from the excited state molecule, one can determine the concentration of a quencher, if its rate constant of interaction with the excited state is known. Once again, consider the molecule, M, that can decay by fluorescence, intersystem crossing or internal conversion. Now add a fourth possibility, quenching. The reactions and rates are then given by:

	Rate
$M \rightarrow {}^1M^*$	I_{abs}
${}^1M^* \xrightarrow{k_f} M + h\nu_f$	$k_f[{}^1M^*]$
${}^1M^* \xrightarrow{k_{isc}} {}^3M^* + \text{heat}$	$k_{isc}[{}^1M^*]$
${}^1M^* \xrightarrow{k_{ic}} M + \text{heat}$	$k_{ic}[{}^1M^*]$
${}^1M^* + Q \xrightarrow{k_q} \text{quenching}$	$k_q[{}^1M^*][Q]$

Note that the quenching reaction rate depends on 2 reactants, M and Q, and the rate constant is a bimolecular rate constant. Once again, one can make the steady state approximation, and compare the rate of fluorescence, kf[1M*] in the presence of the quencher to the same quantity in the absence of the quencher. This leads to the following relationship:

$$\frac{I_f^\circ = 1 + k_q \tau_f [Q]}{I_f} \quad where\ I_f^\circ = k_f \left[{}^1M^* \right]_\circ \ I_f = k_f \left[{}^1M^* \right]$$

The Electromagnetic Spectrum

Photons have an energy which is dependent upon the wavelength of the light. The rule is:

Long wavelength light = low energy

Short wavelength light = high energy

When an opera diva shatters a glass with her singing voice, the glass absorbs resonant sound energy. To be "resonant" the frequency (or wavelength or energy) of the sound must exactly match that of the glass. And so it is with electromagnetic radiation and chemical species.

Spectroscopy is the study of absorption and emission of photons by atoms, ions and molecules. Studies reveal details of atomic, molecular and ionic structure. However, the revealed world is often strange because it is governed by quantum mechanics and subtle selection rules.

Photons of electromagnetic radiation of many wavelengths (or energies) interact with chemical species in a variety of ways.

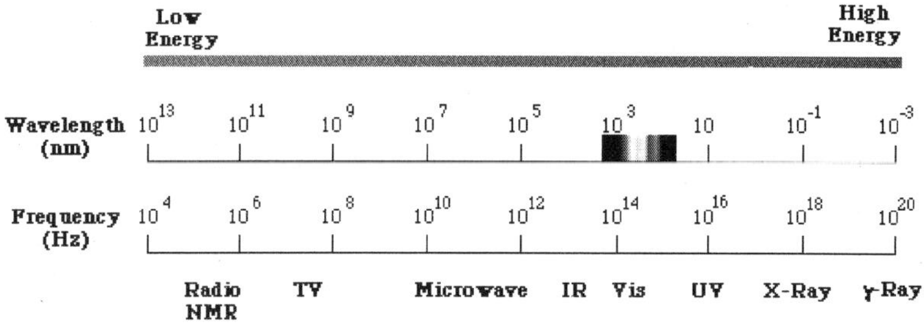

- In a strong magnetic field many atomic nuclei are made to resonate by radio waves. This is the nuclear magnetic resonance (NMR) effect.
- In the gas phase molecules are made to rotate by microwaves.
- Infrared radiation causes molecular bonds to stretch and vibrate. Infrared spectroscopy is often called "vibrational spectroscopy".
- Visible radiation induces low energy electronic transitions in atoms and molecules.
- Ultraviolet radiation causes high energy electronic transitions in molecules. The resulting excited states may relax via bond breakage. Traditionally most photo reaction chemistry has been performed with UV lamp sources. However, laser technology is extending the wavelength range of photoactivation.
- X-rays excite and eject inner shell electrons. This causes widespread ionisation and bond fragmentation.
- Gamma-Rays are highly penetrating ionising radiation which are generated during transitions in atomic nuclei.

Bibliography

Alexandrov, S.: *Methods for Dissolution, Separation and Preconcentration in Analytical Chemistry,* Narodna Kultura, Sofia, 1995.

Arnaudov, M.: *Application of Infrared Spectroscopy in Chemistry,* Nauka i Izkustvo, Sofia, 1981..

Atkins, P.: *Inorganic Chemistry*, London: Blackwell Scientific Publications, 1994.

Balescu, R.: *Equilibrium and Non-equilibrium Statistical Mechanics*, Wiley-Interscience, New York, 1975.

Balram Pani: *Textbook of Environmental Chemistry*, I K International, Delhi, 2007.

Bernard S. Matisoff: *Handbook of Electronics Manufacturing Engineering*, Chapman & Hall, New York, 1997.

Billmeyer, F.W.: *Textbook of Polymer Science,* New York: American University Press, 2001.

Bohren, C.F., and B. Albrecht: *Atmospheric Thermodynamics*. Oxford University Press, 1998.

Bond, A. Curtis: *Modern Polarographic Methods in Analytical Chemistry*, M. Dekker, New York, 1991.

Bontchev, P.R., *Complexation and Catalytic Activity*, Nauka i Izkustvo, Sofia, 1972.

Bower, D. I.: *The Vibrational Spectroscopy of Polymers*, London: Hurst and Company Publishers, 2004.

Bowker, R.: *The Nature of the Chemical Bond*, London: Free Press, 1995.

Bruce, Donald : *Engineering Genesis: The Ethics of Genetic Engineering in Non-Human Animals*, London: Earthscan, 1998.

Campbell, I. M.: *Introduction to Synthetic Polymers, USA*: Oxford University Press, 2002.

Carstensen, J. T.: *Pharmaceutics of Solids and Solid Dosage Forms*, Wiley-Interscience, New York, 1977.

Chambers, K. L.: *Biochemical Coevolution: Eugene*, Oregon State University Press, 1970.

Christian, G. D.: *Analytical Chemistry*, Wiley, NY, 1992.

Cordesman, Anthony H.: *The Gulf and the West: Strategic Relations and Military Realities*, Westview Press, Boulder, Colo., 1988.

Cussler, E. L.: *Diffusion: Mass Transfer in Fluid Systems*, Cambridge University Press, Cambridge, UK, 2009.

Daniels, F., and R. A.: *Alberty, Physical Chemistry*, Wiley, New York, 1955.

Datta-Barua, Lohit: *Natural Gas Measurement and Control: A Guide for Operators and Engineers*, McGraw-Hill Professional Book Group, London, 1991.

De Groot, S.R., Mazur, P.: *Non-equilibrium Thermodynamics*, North-Holland, Amsterdam, 1962.

Dedina, J.; D.L. Tsalev: *Hydride Generation Atomic Absorption Spectrometry*, John Wiley & Sons, Chichester, 1995.

Denbigh, K.: *The Principles of Chemical Equilibrium*, Cambridge Univ. Press, London, 1966.

Donnez, Pierre: *Essentials of Reservoir Engineering*, Editions Technip, Paris, 2007.

Douglas A.: *Principles of Instrumental Analysis*, Thomson Brooks/Cole, Australia, 2007.

Duxbury, G.: *Infrared Vibration-Rotation Spectroscopy, From Free Radicals to the Infrared Sky*, Wiley, NJ, 1999.

Egan, Walter G.: *Photometry and Polarization in Remote Sensing*, New York, 1985.

Ezekwe, Nnaemeka: *Petroleum Reservoir Engineering Practice*, Prentice Hall, Upper Saddle River, NJ, 2011.

Faulkner, L.R.: *Electrochemical Methods: Fundamentals and Applications*, John Wiley & Sons, New York, 2000.

Fraassen, B.: *A Formal Approach to the Philosophy of Science*, Pittsburgh: University of Pittsburgh Press, 2000.

Geis, I.: *Chemistry, Matter, and the Universe*, Menlo Park, Ca., W.A. Benjamin, 1976.

Grandy, W.T., : *Entropy and the Time Evolution of Macroscopic Systems*. Oxford University Press, 2008.

Guyton, A.C.; Hall, J.E.: *Textbook of Medical Physiology*, Saunders, Philadelphia, 2006.

Halliwell, J.J.: *Physical Origins of Time Asymmetry*. Cambridge, 1994.

Herbert, N.: *Quantum Reality: Beyond the New Physics*, New York: Doubleday, 2003.

Hill, T. L.: *An Introduction to Statistical Thermodynamics*, Addison-Wesley, Reading, MA, 1960.

Hines, A. L., and R. M. Maddox: *Mass Transfer: Fundamentals and Applications*, Prentice-Hall PTR, Upper Saddle River, NJ, 1985.

Holland, C. D.: *Fundamentals of Multicomponent Distillation,* McGraw-Hill, New York, 1981.

Holler, F. James: *Principles of Instrumental Analysis,* Brooks/Cole, Pacific Grove, CA, 1998.

Hubert H. Girault: *Analytical and Physical Electrochemistry,* EPFL. , Lausanne, 2004.

Jaynes, E.T.: *Probability Theory: The Logic of Science,* Cambridge University Press, Cambridge, 2003.

Jeliazkova, B. ; V. Michaylova: *Practice in Quantitative Analysis,* University Press, Ohridski", 2000.

Jones, Denny: *Principles and Prevention of Corrosion,* Upper Saddle River, Prentice Hall, New Jersey, 1996.

Joseph, C. Salamone: *Polymeric Materials Encyclopedia,* CRC Press, Boca Raton, FL, 1996.

Kenneth I. Ozomwna: *Recent Advances in Analytical Electrochemistry 2007,* Transworld Research Network, 2007.

Komoroski, R. A.: *High Resolution NMR Spectroscopy of Synthetic Polymers,* Philadelphia: University Press, 2006.

Kondepudi, D.: *Introduction to Modern Thermodynamics,* Wiley, Chichester UK, 2008.

Konopinski, E.J.: *Electromagnetic Fields and Relativistic Particles,* McGraw-Hill, New York, 1981.

Laidler, Keith: *The World of Physical Chemistry,* Oxford University Press, Oxford, 1993.

Lavenda, B.H.: *Thermodynamics of Irreversible Processes,* Macmillan, London, 1978.

Lehninger, Albert: *Principles of Biochemistry,* Worth Publishers, 1993.

Lindon, John C.: *Encyclopedia of Spectroscopy and Specrometry,* Academic Press, Uk, 1999.

Ludwig; Stöcker, Horst: *Thermodynamics and Statistical Mechanics,* Springer-Verlag, UK, 1995.

March, Jerry: *Advanced Organic Chemistry: Reactions, Mechanisms, and Structure,* Wiley, New York, 1985.

Mc. Cash Elaine: *Fundamentals of Molecular Spectroscopy,* Mc Graw-Hill, NY, 1994.

Michael F. Ashby.: *The Materials Selector,* CRC Press, Boca Raton, 1997.

Miller, J. N. and J. C. Miller: *Statistics and Chemometrics for Analytical Chemistry,* Prentice Hall, UK, 2000.

Mitchum, R.M., Jr., Vail, P.R., and Sangree, J.B.: *Stratigraphy – Applications to Hydrocarbon Exploration,* American Association of Petroleum Geologists, Memoir, 1977.

Mitewa, M. : *Department of Analytical Chemistry 1924–1994 - Bibliography*, University Press, Sofia, 1994.

Nicholson, J. W.: *The Chemistry of Polymers,* Oxford: Blackwell Publishers, 2005.

Partington, J.R.: *A Short History of Chemistry*, New York: Dover Publications, 1937.

Popper, K.: *Conjectures and Refutations*, London: Routledge and Kegan Paul, 2005.

Prasad, S.K. : *Biochemistry of Carbohydrates,* Discovery Pub, Delhi, 2010.

Prichard, E.: *Analytical Measurement Terminology*, LGC/Royal Society of Chemistry, Cambridge, UK, 2001.

Rossini, F. D.: *Chemical Thermodynamics*, Wiley, New York, 1950.

Sawyer, L. C.: *Polymer Microscopy*, London, Zed Books Publication, 2004.

Seeman, N.C.: *Biomolecular Stereodynamics*, Adenine Press, New York, 1981.

Sheng, James: *Modern Chemical Enhanced Oil Recovery: Theory and Practice,* Gulf Professional Pub., Amsterdam, Boston, 2011.

Simeonov, V.: *Information Aspects in Chemical Analysis*, Obrazovatelni Tekhnologii, Sofia, 1999.

Skoog, D.A.: *Principles of Instrumental Analysis*, Saunders College Publishing, 1991.

Stone, D. C.: *Instrumental Analysis Laboratory Manual*, Department of Chemistry, University of Toronto, 2005.

Thompson, J. E.: *A Practical Guide to Contemporary Pharmacy Practice*, Williams & Wilkins, Baltimore, 1998.

Tsalev, D.L.; I.P. Havezov: *Atomic Absorption Analysis*, Khimia, Leningrad, 1983.

Van Ness, H.C.: *Understanding Thermodynamics*. Dover Publications, 1983..

Wang, Joseph C.: *Analytical Electrochemistry*, John Wiley & Sons, Chichester, 2000.

Wang, Y.: *Physical Chemistry and Photobiology of Nucleic Acids: Chemistry*, Athens: U of Georgia Press, 2005.

William, L. Leffler: *Petrochemicals in Nontechnical Language,* PennWell, Tulsa, Okla, 2001.

Wolf, P. R.: *Elements of Photogrammetry*, New York, 1983.

Ziegler, H.: *An Introduction to Thermomechanics*, North-Holland, Amsterdam, 1983.

Index

❑❑❑